Recent Titles in
Contributions in Philosophy

Guilt and Gratitude: A Study of the Origins of Contemporary Conscience
Joseph Anthony Amato II

Nature, Human Nature, and Society: Marx, Darwin, Biology, and the
Human Sciences
Paul Heyer

The Meaning of Suffering: An Interpretation of Human Existence from the
Viewpoint of Time
Adrian C. Moulyn

Ethics of Withdrawal of Life-Support Systems: Case Studies on Decision Making
in Intensive Care
Douglas N. Walton

Insight-Imagination: The Emancipation of Thought and the Modern World
Douglas Sloan

Making Believe: Philosophical Reflections on Fiction
C. G. Prado

Arguer's Position: A Pragmatic Study of *Ad Hominem* Attack, Criticism,
Refutation, and Fallacy
Douglas N. Walton

Physician-Patient Decision-Making: A Study in Medical Ethics
Douglas N. Walton

Rethinking How We Age: A New View of the Aging Mind
C. G. Prado

Rationality in Thought and Action
Martin Tamny and K. D. Irani, editors

The Logic of Liberty
G. B. Madison

Coercion and Autonomy: Philosophical Foundations, Issues, and Practices
Alan S. Rosenbaum

EINSTEIN AND THE HUMANITIES

Edited by
DENNIS P. RYAN

Prepared under the auspices of Hofstra University

Contributions in Philosophy, Number 32

Greenwood Press
New York • Westport, Connecticut • London

Library of Congress Cataloging-in-Publication Data

Einstein and the humanities.

(Contributions in philosophy, ISSN 0084-926X ;
no. 32)
Includes index.
1. Einstein, Albert, 1879-1955. 2. Science—
Philosophy. 3. Thought and thinking. I. Ryan, Dennis P., 1946- .
II. Hofstra University. III. Series.
QC16.E5E394 1987   193   86-19444
ISBN 0-313-25380-3 (lib. bdg. : alk. paper)

Library of Congress Catalog Card Number: 86-19444
ISBN: 0-313-25380-3
ISSN: 0084-926X

First published in 1987

Greenwood Press, Inc.
88 Post Road West, Westport, Connecticut 06881

Printed in the United States of America

∞

The paper used in this book complies with the
Permanent Paper Standard issued by the National
Information Standards Organization (Z39.48-1984).

10 9 8 7 6 5 4 3 2 1

Contents

Foreword   xi

Part I   Albert Einstein as a Human Being

1. Einstein and Michelson: Artists in Physics
   Dorothy Michelson Livingston   1

Part II   Ethics and Epistemology

2. The Absolute Beneath the Relative: Reflections on
   Einstein's Theories
   Stanley L. Jaki   5

3. The Moral and Philosophical Origins and Implications
   of Relativity
   Keith R. Burich   19

4. Einstein and Epistemology
   Burton H. Voorhees and Joseph R. Royce   25

Part III   Religion

5. Religion, Relativity, and Common Sense: Einstein
   and the Religious Imagination
   William F. Lawhead   37

6. Einstein on Kant, Religion, Science, and Methodological Unity
   Roy D. Morrison II   47

7. Einstein and African Religion and Philosophy: The
   Hermetic Parallel
   Charles A. Frye   59

Part IV   Metaphysics

8. The Nature of Causality and Reality: A Reconciliation of the
   Ideas of Einstein and Bohr in the Light of Eastern Thought
   Richard Dobrin   73

9. Einstein and the Limits of Reason
   Richard Fleming   81

10. Ontological Relativity: A Metaphysical Critique of
    Einstein's Thought
    Ashok K. Gangadean   87

Part V   History and Philosophy of Science

11. Relativity Before Einstein: Leo Hebraeus and
    Giambattista Vico
    William Melczer   99

12. Einstein, Extensionality, and the Principle of Relativity
    Dennis A. Rohatyn and Patrick J. Hurley   105

Part VI   Literature

13. "Springtime of the Mind": Poetic Responses to
    Einstein and Relativity
    Carol Donley   119

14. The Circuitous Path: Albert Einstein and the Epistemology of
    Fiction
    Robert Hauptman and Irving Hauptman   125

15. A Search for Form: Einstein and the Poetry of Louis Zukofsky
    and William Carlos Williams
    Stephen R. Mandell   135

Part VII   Politics

16. Einstein on War and Peace
    Thomas Renna   143

17. Political Origins and Significance of China's Einstein
    Centennial
    Edward Friedman   151

18. Einstein's Views on Academic Freedom
    Kenneth Fox   165

Part VIII   Education and Psychology

19. Einstein and Psychology: The Genetic Epistemology of
    Relativistic Physics
    Robert I. Reynolds   169
20. Epistemological and Psychological Aspects of Conceptual
    Change: The Case of Learning Special Relativity
    George J. Posner, Kenneth A. Strike, Peter W. Hewson,
    and William A. Gertzog   177

Program of Conference   191
Index   207
About the Editor and Contributors   211

Foreword

Hofstra University is very pleased to present this volume of the proceedings of its Albert Einstein Centennial Conference.

In recent years, Hofstra has become an international conference center attracting scholars from many parts of our nation and abroad. It has been our pleasure to present a series of conferences on subjects which included Feodor Dostoyevski, George Sand, William Cullen Bryant, Franklin Delano Roosevelt, Walt Whitman, Johann Wolfgang von Goethe, and Nineteenth-Century Women Writers. Other conferences deal with Harry S. Truman, Washington Irving, Jose Ortega y Gasset, Dwight David Eisenhower, George Orwell and Twentieth-Century Women Writers.

We are pleased that the scholarly deliberations of the participants at our conferences will be reaching a wider audience through the printed word. Through this volume and its successors, new research will find its way to libraries, scholars and students.

The commitment to scholarship, as typified by the Hofstra conferences, is wholly consistent with the principle that the university should be the arena for ideas.

James M. Shuart
President
Hofstra University

Part I

Albert Einstein as a Human Being

1.
Einstein and Michelson:
Artists in Physics

DOROTHY MICHELSON LIVINGSTON

Albert Einstein and Albert A. Michelson are excellent examples of the complementarity between theorists and experimentalists. In 1879, the year of Einstein's birth, Michelson, at twenty-seven, measured the speed of light which would figure in Einstein's equation $E=mc^2$, and thus their somewhat involuntary association began.

Michelson's attempts from 1880 to 1887 to measure the speed of the earth through the ether played even more closely into Einstein's hands. Michelson and a chemist, Edward Morley, assembled three mirrors and a lens to flash one beam of light moving against the earth's motion and one across this motion. As a ship drags a log in the water to determine its speed, so Michelson hoped to record the earth's speed through the ether.

To his bitter disappointment, the outcome was negative. That is, the two beams of light projected in different directions came together simultaneously, whereas he had hoped that the beam moving against the earth's motion in orbit would lose the race.

News of this negative result spread rapidly and physicists were sorely puzzled. Hendrik A. Lorentz wrote a lengthy paper in 1895, in which he described the Michelson-Morley experiment in detail. The FitzGerald-Lorentz contraction theory explained the negative result by assuming that a moving object was shortened in length by an amount corresponding to its speed in motion. Michelson's interferometer, as his assembled mirrors were called, moving with the earth's speed in orbit, would be compacted just enough to defeat the measurement.

Einstein was sixteen at the time, ten years before he published the special theory of relativity. The angle that stimulated his thinking more than any other was the proof by the ether drift experiments that light moves at exactly the same speed whether it is approaching or receding from the observer. With this information, Einstein quietly declared the ether irrelevant. How simple! No further problem.

Einstein looked for recognition from four men in particular after publication of the special theory: Ernst Mach, his mentor; H. A. Lorentz, with whom he had corresponded; Henri Poincaré, who coined the word "relativity"; and Michelson. To his disappointment none of them came forth

with definite approval. In 1911 Michelson spent a winter in
Gottingen where the younger physicists were full of talk
about Einstein's revolutionary ideas. They gathered after
class in one of the popular bierstuben, dividing themselves
between two tables, one prorelativity and the other against
the new theory. As Michelson appeared after one of his
lectures, everyone watched as he hesitated between the two
and then took his seat decisively among the conservatives.
Eventually, of course, he was forced to accept the new
concept, but it was not without a certain nostalgia for what
he referred to in 1929 as "my beloved ether."

Where these two men differed most was in politics.
Einstein had a world view of civilization as a whole, whereas
Michelson, with his navy background, was passionately
nationalistic. Loyalty to his country, particularly under
attack, took precedence even over his investigations on the
behavior of light. Einstein, on the contrary, was somewhat
removed from the everyday world. My parents found him rather
naive on lay matters. During prohibition he argued for
restoring the saloon, equating it with the German bierstuben,
where intellectuals gather to exchange ideas.

As their friendship grew during Michelson's later years
in Pasadena, they found they shared many tastes. Both of
them enjoyed sailing small boats and both of them played the
violin with considerable skill. They agreed that Mozart was
their favorite composer. Einstein took his violin with him
wherever he went. Michelson found relaxation painting land-
scapes in watercolor. Science came first of course.

The strongest bond existed in their belief in the
importance of an aesthetic approach to their scientific
problems. Addressing his audience at the Lowell Institute in
1899, Michelson said:

> If a poet could at the same time be a physicist,
> he might convey to others the satisfaction, almost
> the reverence which the subject inspires. The
> aesthetic side of the subject is, I confess, by no
> means the least attractive to me. Especially is
> its fascination felt in the branch which deals
> with light.

Many years later after Michelson's death, Einstein paid
him this tribute: "I always think of Michelson as the artist
in science. His greatest joy seemed to come from the experiment itself and the elegance of the method employed."

Part II

Ethics and Epistemology

2.

The Absolute Beneath the Relative: Reflections on Einstein's Theories

STANLEY L. JAKI

Einstein's work on relativity was not yet completed when it began to be taken for the scientific proof of the view that everything is relative. Such a view, widely entertained on the popular as well as on the academic level, is now a climate of thought. A stunning proof of this is a full-page advertisement in the September 24, 1979, issue of _Time_ magazine.[1] It proclaims, under the picture of Einstein, in boldface letters the message: "EVERYTHING IS RELATIVE." The basic rule in advertising, it is well to recall, is a reliance on commonly accepted beliefs; on generally shared cravings, hopes, and fears; or, in short, on the prevailing climate of thought.

The claim that something absolute may be lurking beneath relativity theory may therefore be surprising, though not original at all. That Einstein's relativity theory implies elements and considerations that are absolutist in character was voiced by Max Planck as early as 1924 in the address, "From the Relative to the Absolute," which quickly acquired worldwide publicity.[2] Somewhat earlier Einstein himself began to make statements about the indispensability of metaphysics,[3] This gave no comfort to positivists and empiricists, so many of whom were supporters of the view, in one sense or another, that there is nothing absolute and that, therefore, everything is relative. It could not have, therefore, come as a surprise to Philipp Frank that, as he lectured on relativity at the meeting in Prague of German physicists in 1929, a participant publicly warned him about the absolutist character of Einstein's ideas.[4] Frank refused to take heed for the rest of his life. The main proof of this is Frank's _Relativity: A Richer Truth_, a book distinctly insensitive to the perspective in which Einstein viewed relativity in particular and the philosophy of physics in general.[5]

The essence of that warning given at that congress to Frank was that Einstein fully agreed with Planck that physical laws describe a reality that is independent of the perceiving subject. Doubts on that point were no longer permissible in 1931 when there appeared in print Einstein's contribution to the James Clerk Maxwell commemorative volume, a contribution which began with the famous declaration: "Belief in an external world independent of the perceiving

subject is the basis of all natural science."[6] Twenty years
later, when the Vienna Circle regrouped itself in the United
States, renewed efforts were made by spokesmen of the circle,
such as Hans Reichenbach, to elicit a word or two from
Einstein on behalf of their own "relativist" interpretation
of Einstein's relativity. Einstein did not encourage them,
though being aware that in turn, as he put it, they would
charge him with the "original sin of metaphysics."[7] In his
last essay on relativity, written in 1950, Einstein stated
nothing less than that every true theorist was a tamed meta-
physician, no matter how pure a positivist he fancied himself.[8]

In all these statements Einstein denounced positivism,
endorsed a realist metaphysics, and professed his belief in
the objectivity of physical reality. These and other statements were so many public and emphatic indications of his
belief that there was something absolute beneath the
relative. Yet, one would look in vain for a substantive
trace of those statements in the books and articles written
on relativity by Frank, Rudolf Carnap, Reichenbach, and
Herbert Feigl, all members of the Vienna Circle, who in the
1950s and 1960s captured, in the United States at least, the
role of authoritative spokesmen on behalf of Einstein in
particular and of science in general. Their systematic
silence on many relevant statements and facts was only part
of the strategy pursued by them. Instead of strategy, the
word crusade would be more appropriate. Reichenbach himself
warned that logical positivism should be looked upon as a
crusade and not as an abstract philosophizing.[9] Intellectual
crusades have their inner logic to which logical positivists
were not immune. Or, as Herbert Feigl admitted well over a
decade ago: "Confession, it is said, is good for the soul.
Undoubtedly we [logical positivists] made up some facts of
scientific history to suit our theories."[10]

Such a confession, rather incriminating for positivists, logical or other, who profess to be respectful only
of facts, is hardly a spontaneous one. It is most likely
triggered when a carefully contrived and nurtured make-
believe or illusion is suddenly punctured. As to the long-
cherished balloon of Einstein's positivism, it received a
particularly stinging blow through the publication, in 1968
and 1969, respectively, of two extensive studies by Gerald
Holton, "Mach, Einstein, and the Search for Reality," and
"Einstein, Michelson, and the Crucial Experiment."[11] Neither
of these massively documented essays is without some short-
comings. Although in Einstein's formulation of special relativity the experiment of Albert Michelson and other experiments devised for the detection of the ether played no
crucial role, they were familiar to Einstein and played some
role in his reasoning. As to Einstein's departure from and
opposition to Ernst Mach concerning reality, Holton did not
quote two letters of Einstein which are particularly ex-
pressive in this respect and will be discussed later.

It would be rather naive to assume that such and
similar documentations and Einstein's own statements,repeated
over four decades, would be effective in discrediting the
climate of thought in which an allegedly exclusive respect
for facts supports the tenet according to which everything is
relative, and especially that all values are relative. The

ludicrous worshiping of "facts alone" and its invitation to
unabashed selfishness, if not dishonesty, once the concomitant relativization of values is made full advantage of, were
already immortalized in Dickens' Hard Times. Clearly, the
climate of thought in question had existed long before
Einstein's relativity was cited on its behalf. Of the long
story of the relativization of truth and values in Western
thought, let it suffice here to note that pragmatism and
behaviorism were catchwords for a long time before it became
fashionable to justify them with copious references to a
theory of physics, known as relativity.[12] A striking
illustration of the relativization of truth and values as it
asserts itself in our own days is that "crazy quilt of revised judgements" — the concise summary by an anonymous
reviewer of the picture which emerges from a recent survey of
textbooks on American history.[13] Not that Frances
Fitzgerald, the author of that survey, is particularly happy
with the systematic discrediting of traditionally shared
views on the foundation and purpose of this nation of ours.
But she offers a very revealing justification of this
unpleasant process: "All of us children of the 20th century
know or should know that there are no absolutes in human
affairs." She also specifies the source of this knowledge as
"the pluralism or relativisation of values."[14]

It is a redeeming value of her reasoning that she does
not invoke Einstein's relativity as a support, a surprising
departure from a standard technique. That the technique is
such a standard can be gathered from the advertisement in
Time which also offered as an unquestioned verity that "In
the cool, beautiful language of mathematics Einstein demonstrated that we live in a world of relative values." The
statement is as misleading as almost anything that makes for
a flashy advertisement, but, as all such advertisements, it
reflects a tone of thought, or at least an
unconscious wishful thinking — otherwise it would not have
been seized upon by a highly professional advertising
agency.[15] Interested in quick profit, such agencies are not
the ones to ask whether indeed Einstein had ever tried to
prove in the language of mathematics, or in any language,
that all values are relative.

To find the answer to this question a few hours of
reading of Einstein's essays, or a consultation with anyone
familiar with his writings and not blinded by positivism,
would have been sufficient. Einstein never tried such a
demonstration and certainly not in the cool and beautiful
language of mathematics. This is not to suggest that
Einstein offered no clues to his own thinking about values or
that he was original or consistent in this respect. He
merely voiced an old cliche when, in the foreword which he
volunteered to Frank's Relativity: A Richer Truth, he specified man's instinctive avoidance of pain as the source of
value judgments and of ethics itself. On this basis the
relativity of values could only be a foregone conclusion.[16]
It is, of course, well known that for all his dismissal of
religion and of belief in a personal God, Einstein insisted
on the unquestionable superiority of the Judeo-Christian perspective in which unconditional value is attributed to each
and every human being. But his insistence was incompatible
with mechanistic evolutionism which he also endorsed,

although it provides, as had already been pointed out by such
a protagonist of Darwin as T. H. Huxley, no room for "higher"
and "lower."[17]

To his credit, Einstein consistently avoided basing his
views on values and ethics on his theory of relativity and on
mathematics. This shows something of his instinctive great-
ness, because history knows of some misguided men of science
(Maupertuis, and Condorcet, for instance) who tried to con-
strue ethical theories from manipulating quantities.[18] As to
his own theories, which, as will be seen, were more than mere
mathematics, he stated emphatically four years before his
death: "I have never obtained any ethical values from my
scientific work."[19] To be sure, he made a few memorable
utterances concerning freedom and oppression, but his general
trend was to avoid involvement in human affairs. He declined
the presidency of Israel with a reference to his lack of
familiarity with human nature and affairs. Tellingly, his
autobiography opens with the remark that he had never regret-
ted that he had left behind the customary human world and
moved into the strange, depersonalized world of science.[20]

Clearly, "the absolute beneath the relative" should in
connection with Einstein's theories be sought in a direction
different from what leads to values and ethics. Of the three
main theories of Einstein — special relativity, general
relativity, and unified field theory — the first, on a
cursory look at least, does not give a clue as to what the
direction might be. The article in which Einstein formulated
special relativity in 1905 has become the victim of a
stereotyped reading. In the crudely superficial version of
that reading, special relativity is an effort to explain the
Michelson-Morley experiment. According to the moderately
superficial version, special relativity "has its roots in the
questions: Where are we? How are we moving?" An example of
this latter version is the article "Relativity" by Banesh
Hoffmann in the Dictionary of the History of Ideas, an
article which starts with the foregoing questions.[21] Both
readings can claim for their support one and the same phrase
which, after a reference to electromagnetic induction and to
the unsuccessful attempts to discover any motion of the earth
relative to the ether suggests that "the phenomena of electrodynamics as well as of mechanics possess no properties
corresponding to the idea of absolute rest."[22] However, the
explanation of the unsuccessful attempts had already been
given by the Lorentz transforms and by the contraction postulated by George FitzGerald. As to the absolute rest, its
critique had already been offered two hundred years earlier
by Berkeley on purely kinematic grounds. There had to be
some specific and novel rationale in Einstein's handling two
well-worn topics. The clue of this is given in the phrase
which immediately follows the one just quoted above. There
Einstein goes beyond the question of absolute rest with the
remark that the null results of those experiments rather
suggest that "the same laws of electrodynamics and optics
will be valid for all frames of reference for which the
equations of mechanics hold good."

In itself the phrase is rather ambiguous: in the light
of Einstein's train of thought leading to general relativity
and to unified field theory, the phrase is a classic of the
inability of a genius to say explicitly what was truly in the

back of his mind. Had Einstein italicized the word <u>same</u>, he
would have strongly intimated that his principal concern was
neither the explanation of the Michelson-Morley experiment,
nor the problematic character of absolute rest.   It was
rather the <u>sameness</u> of the laws of electrodynamics, which the
opening phrase of Einstein pointedly introduced as "Maxwell's
electrodynamics."   This electrodynamics had a special place
in Einstein's thought.   In his autobiography he referred to
it as the "most fascinating subject" available in his student
days.[23] Actually, he viewed it as the most fundamental subject in physics.   The proof of this is his contribution in
1931 to the volume commemorating the centenary of Maxwell's
birth.   There, in surveying the latest developments of physical theory, including quantum mechanics, he claimed it as a
certainty that ultimately physics will return to carrying out
"the   program   which   may   properly   be   called   as   the
Maxwellian — namely, the description of physical reality in
terms of fields, which satisfy partial differential equations
without singularities."[24]

The singularities implied by the context were the
material points (particles) which in Newton's physics represented the bedrock of reality.   They were replaced by fields
in Maxwell's theory which, of course, did not mean the elimi-
nation of such singularities as constants and boundary conditions.   But the notion of a field could not function as the
post-Newtonian foundation of physics if it was the function
of a particular frame of reference.   Its independence of any
frame of reference could only be safeguarded if Maxwell's
equations retained the same form regardless of the frame of
reference in consideration.   This, however, implied the postulate of the constancy of the speed of light regardless of
the motion of its source.   Such is the ultimate justification
of that postulate about which Einstein felt it necessary to
note in his 1905 paper that it was "only apparently irreconcilable with the former" principle, which he unfortunately
labeled "Principle of Relativity."   The label, perhaps the
most unfortunate in the entire history of physics, made him
oblivious to the fact that he failed to reconcile fully two
apparently contradictory points.   One was the principle itself, which on a cursory look stated the relativity of all
positions and motions.   The other was the speed of light,
endowed, as being not relative to the motion of its source,
with an absolute character.   His claim that between these two
points there was no basic irreconcilability made sense only
if the expression "same laws of electrodynamics" meant the
sameness of these laws in a somewhat different but certainly
far deeper sense.   He should have spelled out that if those
laws retained their original form regardless of the frame of
reference to which they were related, it was only because
they reflected an objective, invariant, absolute cosmic order
and reality.

Such   was   the   gist   of   Einstein's   explanation   of
Lorentz's equations, which had already explained the null
result of the Michelson-Morley experiment, but through which
(and this was the all-important point not emphasized by
Lorentz) Maxwell's equations retained the same form even when
related to a frame of reference which moved at constant
velocity with respect to another.   That in 1905 Einstein
himself was not entirely clear or explicit as to what was the

real driving force behind his reasoning is a secondary matter.[25] The important thing is that his mind was in the grip of that driving force. It was the attractiveness of a specific vision of nature and of a most fruitful scientific interpretation of it.

The vision was that of a cosmic reality, fully coherent, unified, and simple, existing independently of the observer; that is, not relative to him, and yielding its secrets in the measure in which the mathematical formulas, through which it was investigated, embodied unifying power and simplicity. In the case of special relativity there was already a most unexpected and unintended yield: the absolute energy content of a mass at rest, given in the now historic formula $E=mc^2$.[26] Although at that time experimental evidence on behalf of that formula was ambiguous, Einstein upheld its validity by referring to the broad theoretical foundation on which it rested. The foundation was much broader than it appeared to be. The proof of this is his first essay on general relativity, running over fifty pages, which was already in print in 1907.[27] Clearly, if special relativity had not been far more than the explanation of the null result of the Michelson-Morley experiment and an answer to the questions of where we are and how we move, Einstein would not have faced up to the problems of general relativity while the printer's ink was still fresh on his special relativity. His real concern was the elaboration of a cosmic view in which physical reality was a totality of consistently interacting things, an absolute in the sense that its existence was not relative to any observer, and absolute also in the sense that if the observer's knowledge of reality was properly scientific, the laws in question had to remain as invariant as the universe is invariant. Indeed, Einstein himself suggested that special relativity should have been called the theory of invariance.

On the face of it, general relativity is a further exercise in relativization. The impossibility of specifying any frame of reference as privileged over any other that move with respect to one another with constant velocity is extended in general relativity to all frames of reference that are accelerated with respect to one another. The three classic observational consequences of general relativity (the gravitational red shift, the gravitational bending of light, and the precession of the perihelion of planetary orbits, observable in the case of Mercury) implied not only relativization but also equivalence or unification, namely, the equivalence of gravitational and inertial masses. That the thrust of general relativity was indeed unification became all too obvious with the appearance in 1917 of the paper, "Cosmological Considerations on the General Theory of Relativity."[28] Instead of "considerations" Einstein should have perhaps written "consideration." The considerations he specified (the value of the total mass of the gravitationally interacting matter, the value of the radius of that totality, or the universe, the curvature of space-time) are well known. What is hardly ever recalled is the fact that all such considerations rest on one basic consideration: the power or ability of general relativity to treat in a scientifically consistent manner the totality of material particles endowed with gravitation. That ability made scientific cosmology possible for the first time.

There were, of course, cosmologies before Einstein. Their scientific insignificance is not primarily the outcome of the relatively meager data that were available about the cosmos prior to the twentieth century. What makes those pre-Einstein cosmologies scientifically insignificant is that (with the exception of one proposed by Johann Lambert) they were not free of a basic theoretical defect of which there was a sufficient awareness already in Newton's time. The defect concerns the infinity paradox which plagued the notion of the idea of an infinite universe whether it was homogeneous or hierarchical. The idea of a homogeneous infinite universe is usually connected with Newton's name. The basis of this connection is that the idea began to be mentioned by some scientists only from his time on. Although Newton, as it appears from his letters to Richard Bentley, seemed to think that an infinite homogeneous universe of stars is gravitationally possible, he never departed from his early belief that the universe is finite whereas space itself is infinite.[29] Indeed, no protest was heard either from Newton or from others when in 1714 Joseph Addison attributed to Newton this idea of a finite universe in infinite space and praised it as the notion most worthy of reason and of God. Addison did so in the pages of the Spectator which was read all over Europe.[30]

Contrary to cliches in most histories of cosmology and science, the finiteness of the universe was the prevailing view until the early part of the nineteenth century. But as Lambert already pointed out in 1761 such a finite universe had to collapse gravitationally unless all its parts revolved around a center, possibly an enormously massive body. The rotating finite universe proposed by Lambert was hierarchically organized,[31] an organization which had already been proposed by Kant a few years earlier who argued the infinity of a hierarchically organized universe without realizing that his universe had to have an infinitely massive body at its center.[32] Earlier, Edmund Halley tried to save the infinity of the universe by suggesting that the distribution of stars was not homogeneous.[33] He considered only the optical problem but not the gravitational one. In 1823 Heinrich Olbers notoriously failed to consider the gravitational paradox as he tried to solve the optical paradox by a recourse to the absorption of starlight in interstellar space,[34] a procedure already suggested in 1730 by Nicolaas Hartsoeker,[35] and in 1743 by Jean Cheseaux.[36]

There was no echo when in 1872 Johann Zöllner showed both that an infinite homogenous universe was contradictory and that the only consistent way of treating the totality of gravitationally interacting matter was to take it to be finite in a four-dimensional non-Euclidean space. No major discussion followed when in 1895 Hugo Seeliger suggested a change in the inverse square law of gravitation to avoid the gravitational contradiction which arises in an infinite homogeneous universe. Needless to say, the slightest change in the inverse square law made impossible the explanation of planetary motions. In 1901 Kelvin summed up the paradox of an infinite universe in a concise formula, but he skirted the gravitational aspect and solved the optical aspect on the ground that the light coming from beyond the Milky Way was

wholly negligible.[37]   No discussion ensued when C.V.L.
Charlier tried to save infinity in 1911 by assigning a hierarchical structure to the universe.

What these glimpses into pre-Einsteinian cosmology
should suggest is that glaringly defective arguments were
taken in stride as long as they were proposed in defense of
infinity of matter or space or both. Clearly, behind this
nonscientific attitude there must have been some nonscientific motivations. They derived from the fact that it was
tempting to take infinite homogeneity as a necessary form of
existence, that is, something which explained itself and was
its own sufficient raison d'être. The shock therefore was
considerable when in 1922 Einstein emphatically argued at the
Sorbonne on behalf of the finiteness of the total mass of the
universe.[38] Further refinements of estimates of the average
density of matter, which calls for that finiteness of the
total mass, did not fail to corroborate Einstein's argument.
Einstein, of course, was fully aware that it was possible to
construct four-dimensional world models that could accommodate an infinite amount of matter, and even with a
homogeneous distribution. Einstein, however, brushed aside
these models as insignificant, although he himself devised
one, according to which the world lines were helically
cylindrical.[39]

A universe embodying three-dimensional Euclidean homogeneity appears so natural to perception as to be taken for a
natural or necessary form of material existence. A universe
resembling either a spherical cylinder or hyperbolic surface,
open-ended as they could be, must strike one as very specific
and hardly a necessary form of existence. When faced with
such a singular form of existence, one can hardly avoid
facing up to the question: What makes the universe so specific? Of course, the universe need not be cylindrical in
order to prompt this question. It is enough to think of the
value of the space-time curvature which the universe actually
has. It is a strange specific number, different from 0,
which is the curvature of the intrinsically impossible homogeneous Euclidean universe. This 0 is a symbolic indication
that such a universe, like 0, is a figment of imagination,
bordering on mere nothing. A positive number, such as 0.8 or
1.6, standing for the space-time curvature, must strike one
very differently. Looking at such a curvature should do what
is done by a look at the tag of a dress, a tag carrying the
measurement and price of the dress. Such a tag cannot help
evoke the existence of a dressmaker.

Einstein himself was prompted to such considerations.
His general relativity, as the first consistently scientific
treatment of the universe as the totality of gravitationally
interacting entities, reassured him in his previous instinctive conviction that the universe was real and fully
rational. This was one of the reasons why he rejected the
philosophy of Kant for whom the notion of the universe was
merely a bastard product of the metaphysical cravings of the
intellect. Once the notion of the universe was made out to
be intrinsically unreliable, Kant could argue that any step
from the universe to the Creator was also unreliable. But
once the notion of the universe was fully vindicated by
general relativity, Kant's argument and his whole criticism
of natural theology lost whatever credibility it could

marshal.[40] Einstein was most conscious of the full force of this implication. In a letter written four years before his death to his lifelong friend Maurice Solovine, Einstein insisted that it was not permissible to go beyond the universe to its Creator. The letter was a reassurance given by Einstein to Solovine that Einstein, the cosmologist, had not become a believer in a personal God and Creator. He foresaw that his cosmology would be exploited by priests and theologians. "It cannot be helped," Einstein wrote to Solovine. "I add this," Einstein continued, "lest you think that weakened by old age I have fallen into the hands of priests."[41]

Once the universe as a totality of consistently interacting things is recognized as such, all efforts to relativize everything reveal their futility at once. Tellingly, the most convincing proof of that totality, the 3°K cosmic background radiation, has reminded many experts on relativity that the expansion of the universe was a nonrelativist frame of reference.[42] At any rate, if not priests, at least some basic aspects of their theology, must have been in the back of Einstein's mind for a reason relating to his efforts to work out a unified field theory.[43] Twice, in the late 1920s and late 1940s, Einstein thought that he had achieved his goal. As is well known, he failed in both cases. But even if he had succeeded, only gravitation and electromagnetism would have been united and only on the macroscopic level. He did not think that relativity and quantum theory could be united, except by replacing Quantum Theory with something else. He never worked on nuclear forces and was dead by the time the so-called weak forces came to be widely recognized. But with this unified field theory he made a most notable effort toward a goal which has lately excited a special fascination for cosmologists.[44] The goal is the demonstration on theoretical grounds (mathematical and philosophical) that the universe (from atoms to galaxies) can only be what it is and nothing else. Einstein himself dreamed of a unified theory so simple that even the good Lord would not have been able to fashion the world along any other lines.

To his credit, Einstein never entirely parted with the humble recognition that the ultimate word in science belongs to facts, that is, to the observational verification of theories. Indeed, in 1920, he did say that if only one of the three classic proofs of general relativity were to be disproved, all general relativity would turn into "mere dust and ashes."[45] Others, Sir Arthur Eddington for instance, who were animated by the vision of a final theory, were not so mindful of the primacy of facts. A scientist is hardly mindful of facts when he declares before an audience of three thousand that within a few years, but certainly sooner or later, he or others will come up with a theory which shows why the family of elementary particles and therefore the universe can only be what it is and nothing else.[46] A mere recall of the fact that science can never be sure that it knows all the facts should suffice to dispose of such a brash dream. The intrinsic merits of the goal of devising an ultimate theory should also seem nil as long as the theory is sufficiently mathematical, and clearly such a theory must be highly mathematical. Kurt Gödel's incompleteness theorem states that the proof of consistency of any nontrivial set of mathematical axioms can be found only outside that set, and in that sense no mathematical system can be an ultimate system. In other words, whereas general relativity forces us to admit the realistic character of the notion of consistently interacting things, as valid object of scientific cosmology, the application of Godel's theorem to cosmology shows that a disproof of the contingency of the universe is impossible. The mental road to the extracosmic Absolute remains therefore fully open.

These points have been repeatedly made in several of my publications since 1966.[47] Apparently some in the scientific and philosophical community want to learn only what they want to hear, and therefore choose to ignore the tie between Godel's theorem and cosmology. It is, of course, no surprise to me that the contingency of the universe is not pleasant news to a scientific humanism which claims that man is a mere accident, in no way subject to something transcendental to the entire universe. Such a humanism is more powerful in our times than it has ever been. This is why _Time_ felt it natural to proclaim under Einstein's picture that everything is relative. The only message befitting Einstein's picture would have been a warning that the absolute is lurking everywhere beneath the relative. But Time is very human and so are our times, indeed all times. To this rather defective humanness of the times, proper reflections on Einstein's work may bring a much needed corrective. Failing that, there will be no slowing down of that culturally destructive merry-go-round that witnesses the absolutization of the relative by those who are busy relativizing the absolute.

Notes

1. _Time_, September 24, 1979, p. 64. The advertisement, facing p. 64, was on behalf of _Time_ itself.
2. The address, "Vom Relativen zum Absoluten," has been a part of the best known collection of Planck's addresses, Wege zur physikalischen Erkenntnis: Reden and Vorträge, from its first edition (1933) on. A somewhat free English version is available in M. Planck, Where Is Science Going? (New York: W. W. Norton, 1932), pp. 170 - 200. In that address Planck emphasized the absolute value of energy in terms of the formula $E=mc^2$ and the independence of the total four-dimensional space-time manifold from the observer.
3. See G. Holton, "Mach, Einstein, and the Search for Reality" (1968), in Thematic Origins of Scientific Thought (Cambridge, Mass.: Harvard University Press, 1973), p. 243.
4. Reported by Phillip Frank himself in his Einstein: His Life and Times (New York: A. Knopf, 1947), p. 215.
5. Phillip Frank, Relativity: a Richer Truth (London: Jonathan Cape, 1951). The book certainly reveals the futility of the efforts of a pragmatist to vindicate universal validity for democratic way of life on the basis of the "relativity of knowledge." The offsprings of that relativity were, according to Frank, "not only modern science, but also liberal Christianity and reformed Judaism" (p. 20), a statement which gives away its true character to anyone mindful of the chronic inability of both liberal Christianity and of Reform Judaism to proclaim anything absolute.

6.   A. Einstein, The World As I See It (New York: Covici and Friede, 1934), p. 60.
7.   See his "Reply to Criticisms," in Albert Einstein: Philosopher-Scientist, ed. P.A. Schilpp (Evanston, Ill.: Library of Living Philosophers, 1949), p. 673.
8.   Albert Einstein, "On the Generalized Theory of Gravitation," in Ideas and Opinions (New York: Crown, 1954), p. 342.
9.   "The whole movement of scientific philosophy is a crusade. . . . . What I'm doing aims as directly at social consequences as the programs of those who call themselves 'social reformers'." Statement reported by C. Schuster in M. Reichenbach and R. S. Cohen, eds., Hans Reichenbach: Selected Writings, 1909 - 1953 (Dordrecht: D. Reidel, 1978), 1:56 - 57.
10.   H. Feigl, "Beyond Peaceful Coexistence," Historical and Philosophical Perspectives of Science, ed. R. H. Steuwer (Minneapolis: University of Minnesota Press, 1970), p. 3.
11.   The latter, too, is reprinted in Holton's Thematic Origins of Scientific Thought (Cambridge, Mass.: Harvard University Press, 1973), pp. 261 - 352.
12.   Relativity was still a novelty for many physicists in the 1920s which saw the publication of A.Bentley's Relativity in Man and Society (New York: G. P. Putnam's Sons, 1926), with its second chapter entitled: "The Term 'Einstein' — Its Meanings." The next two books, though separated by a World War, a continent, and opposite theses, have one small but revealing detail in common. In his In Quest of Morals (Stanford University Press, 1941), H. Lanz quotes (p. 19) H. Weyl on behalf of his claim that relativity supports philosophical and ethical relativism, who is also quoted, but in the opposite sense, in H. Wein's Das Problem des Relativismus (Berlin: W. De Gruyter, 1950), p. 26. No relevance is accorded to relativity (and no mention of Einstein is made) in three epistemological rebuttals of relativism: G. Rabeau, Réalité et relativité (Paris: Marcel Riviere, 1927); H. Spiegelberg, Antirelativismus (Zurich: Max Niehans, 1935); G. D. Kaufmann Relativism, Knowledge, and Faith(Chicago: University of Chicago Press, 1960). The following two books are relevant also because of the title of this essay. In Relativisme (Paris: Kra, 1930), a little known work by A. Maurois, there is a chapter, "L'Absolu dans le relatif" (pp. 69 - 76), devoted to the impossibility of "complete" relativism, though with no reference to Einstein or relativity. Both Einstein and relativity are prominently in view from the very start in Il n'y a d'absolu que dans le relatif (Paris: J. Vrin, 1975) by R. Levi.
13.   Time, September 10, 1979, p. 68, in a review of America Revised: History Schoolbooks in the Twentieth Century (New York: Little, 1979) by Frances Fitzgerald.
14.   Ibid., p. 69.
15.   That such is the case is palpably shown by another full-page advertisement on behalf of Time, November 12, 1979, p. [124] which, under the picture of two famous ballet dancers, carries the caption: "News, like beauty, is often in the eye of the beholder."
16.   Einstein was not unaware of this possibility, but

to make matters worse, he tried to save the norms of ethics from pure arbitrariness with a reference to "the psychological and genetic point of view" (Frank, Relativity: A Richer Truth, p. 10). In doing so he only presented himself as an easy target to any skillful debater who has been granted the basic philosophical premises of Darwinism.

17. See, More Letters of Charles Darwin: A Record of His Work in a Series of Hitherto Unpublished Letters, ed. F. Darwin and A. C. Seward (New York: D. Appleton, 1903), 1:360.

18. For details, see my The Relevance of Physics (Chicago: University of Chicago Press, 1966), pp. 376 - 78.

19. Quoted in P. Michelmore, Einstein: Profile of the Man (New York: Dodd, 1962), p. 251.

20. Albert Einstein: Philosopher-Scientist, p. 5.

21. B. Hoffman, "Relativity," in Dictionary of the History of Ideas. (New York: Charles Scribner and Sons, 1968 - 1974), 3:74.

22. "On the Electrodynamics of Moving Bodies," in The Principle of Relativity: A Collection of Original Memoirs on the Special and General Theory of Relativity by H. A. Lorentz, A. Einstein, H. Minkowski and H. Weyl, with notes by A. Sommerfield, translated by W. Perrett and G. B. Jeffrey (1923; New York: Dover, n.d.), pp. 37 - 38.

23. Albert Einstein: Philosopher-Scientist, p. 33.

24. A. Einstein, "Clerk Maxwell's Influence on the Evolution of the Idea of Physical Reality," in The World As I See It, p. 66.

25. Inattention to this point lies at the root of that controversial chapter on relativity in E. T. Whittaker's A History of the Theories of Aether and Electricity, Volume Two: The Modern Theories 1900-1926 (London: Thomas Nelson, 1953), pp. 27 - 77, in which Einstein appears a minor figure in comparison with Poincaré and Lorentz.

26. After writing in 1905 the energy content of a mass as being equal to L/c^2 and in 1906 as E/V^2, he finally put in 1907 the energy E as being equal to μc^2, still not exactly the now standard notation.

27. "Ueber das Relativitatsprinzip und die aus dem selben gezogenen Folgerungen," Jahrbuch der Radioaktivitat und Elektronik 4(1907):411 - 62.

28. In The Principle of Relativity, pp. 177 - 88.

29. See A. R. and M. B. Hall, Unpublished Scientific Papers of Isaac Newton (Cambridge: University Press, 1962), p. 138.

30. See, Spectator, July 9, 1714. An equally important witness is Voltaire in the uncounted editions of his Elemens de la philosophie de Newton (1738) following its enlargement with his booklet, La metaphysique de Newton ou parallele des sentimens de Newton et de Leibniz (Amsterdam: chez Jacques Desbordes, 1740), in which he emphasized the infinity of space and the finiteness of matter with a reference to the authority of Newton (p. 2).

31. Lambert did so in his Cosmologische Briefe. See my translation, Cosmological Letters on the Arrangement of the World-Edifice, with an introduction and notes (New York: Science History Publications, 1976).

32. For details see the introduction of my translation

of his Allgemeine Naturgeschichte und Theorie des Himmels, or Universal Natural History and Theory of the Heavens (Edinburgh: Scottish Academic Press, 1981).

33. For a reprint of his two papers, see my The Paradox of Olbers' Paradox (New York: Herder and Herder, 1969), pp. 249 - 52.

34. For a discussion and a reprint of his paper, see Paradox of Olber's Paradox, pp. 131 - 43 and 256 - 64.

35. He did so in his Cours de physique (The Hague: chez Jan Swart, 1730), p. 235.

36. For a reprint of Cheseaux's paper, see my Paradox of Olbers' Paradox, pp. 253 - 55.

37. See my "Das Gravitations-Paradoxon des unendlichen Universums," Sudhoffs Archiv 63 (1979): 105 - 22, and my The Milky Way: An Elusive Road for Science (New York: Science History Publications, 1972), pp. 275 - 77.

38. Typically, the French physicist, E. Borel, was willing to grant only the "convenience" of the finiteness of mass in his exposition of Einstein's theories, L'espace et le temps (Paris: F. Alcan, 1922). See its English translation, Space and Time (London: Blackie and Son, 1926), pp. 226 - 27.

39. "Cosmological Considerations on the General Theory of Relativity," in The Principle of Relativity, p. 179. It should be revealing that Eddington, in his Space, Time and Gravitation: An Outline of the General Relativity Theory (Cambridge: University Press, 1920) objected to Einstein's world-model on the ground that it reinstated absolute space-time (see p. 162)!

40. For the inconclusiveness of Kant's criticism of the cosmological argument, see my The Road of Science and the Ways to God (Gifford Lectures, Edinburgh, 1974 - 75 and 1975 - 76; Chicago: University of Chicago Press, 1978), pp. 121 - 22 and 379 - 80.

41. Letter of March 30, 1952, in A. Einstein, Lettres à Maurice Solovine, reproduites en facsimile et traduites en français (Paris: Gauthier-Villars, 1956), pp. 114 - 15. For longer excerpts in English translation from this and Einstein's preceding letter to Solovine, see my Cosmos and Creator, (Edinburgh: Scottish Academic Press; Chicago: Regnery-Gateway, 1980), pp. 52 - 53.

42. See P. G. Bergmann, "Cosmology as a Science," in Philosophical Foundations of Science, R. J. Seeger and R. S. Cohen, eds. (Dordrecht: D. Reidel, 1974), pp. 181 - 88, who speaks of the "breakdown of the principle of relativity with respect to the background radiation" (p. 185).

43. See my Cosmos and Creator.

44. "This point [the moment of the Big Bang] is thus de facto preferred. . . . Naturally this does not constitute a disproof, but the circumstance irritates me," wrote Einstein to De Sitter on June 22, 1917. See Nature, October 9, 1975, p. 454.

45. A statement made by Einstein in 1920 during a lecture given in Prague which was attended by young H. Feigl who recalled it many years later. See his Historical and Philosophical Perspectives of Science, p. 9.

46. Professor Murray Gell-Mann, at the Twelfth Nobel Conference, October 6, 1976, held at Gustavus Adolphus

18 Stanley L. Jaki

College, St. Peter, Minnesota.
    47.   Such as The Relevance of Physics, pp. 128 - 30;
The Road of Science and the Ways to God, p. 456; Cosmos and
Creator, pp. 49 - 50; and "From Scientific Cosmology to a
Created Universe," <u>Irish Astronomical Journal</u> 15 (March
1982): pp. 253 - 62.

3.

The Moral and Philosophical Origins and Implications of Relativity

KEITH R. BURICH

Despite his obvious ideological bias, it was Lenin who probably possessed the most perceptive understanding of the moral and philosophical underpinnings of relativity. Einstein believed that he had moved one step closer to the ultimate unveiling of a higher, more transcendent set of absolutes which existed independently of the individual or human psyche, and were as valid in the spiritual realm as they were in nature. It was these elements of idealism in relativity which so alarmed Lenin, and which were to become even more pronounced in Einstein's later years as his attention shifted away from science and toward the burning issues of world government, disarmament, and Zionism.

According to Ernst Mach, science was guilty in its metaphysical tendencies of the same egocentrism or anthropomorphism which, in the past, had led to "numerous religious, ascetic and philosophical absurdities."[1] The arbitrary authorities which ideas, including scientific concepts, often hold in men's lives was unwarranted because nature itself possessed no hierarchies, no values, no direction or purpose. And, once these ideas or beliefs hardened into dogma, orthodoxy had to be defended, often at a terrible cost to man and society. Mach's solution was to liberate man from the anthropocentric belief that the laws of nature and society must conform to human ideals. The ego was Mach's original sin, for it deluded man into trusting reason rather than his senses. However, only the senses could emancipate mankind from its self-destructive egotism. Admittedly, Mach's phenomenalism circumscribed the creative potential of the scientist, but many young socialists and Marxists at the turn of the century became enamored with Machian philosophy because his "new" physics challenged any and all claims to absolute authority and pointed the way toward a more objective, more "scientific" synthesis.

Lenin, however, was far more skeptical of Mach's purported materialism. It may have superficially resembled the dialectical materialism of Marx and Engels, but Lenin shrewdly perceived that Mach's denial of the existence of a hidden reality or Ding an Sich independent of individual experience was solipsistic.

No evasions, no sophisms...can remove the clear and

indisputable fact that Ernst Mach's doctrine of
things as complexes of sensations is subjective
idealism and a simple rehash of Berkeleianism. If
bodies are "complexes of sensations," as Mach said,
or "combinations of sensations" as Berkeley said,
it inevitably follows that the whole world is my
idea. Starting from such a premise it is impossible to arrive at the existence of other people
besides oneself: it is the purest solipsism.[2]

Lenin was equally critical of Einstein, and his attacks on
Machian and Einsteinian relativity were repeated in Russian
classrooms and textbooks during the first years of the Soviet
regime.[3]

Actually, Lenin was not entirely unjustified in associ-
ating Einstein with Mach. In his "Autobiographical Notes,"
Einstein himself acknowledged his indebtedness to Mach for
shaking his "dogmatic faith" in the concepts of classical
physics.[4] Just how loyal Einstein was to Machian epistemology in the formulation of his relativity theories remains
problematic. There can be no doubt that the skepticism made
fashionable by Mach, Poincairé, and other "relativists" at
the end of the nineteenth century gave Einstein courage to go
beyond the mechanical concepts of classical physics. However, Einstein soon began to chaff at the heuristic restric-
tions of Machian positivism. As he wrote in his "Autobiographical Notes":

By and by I despaired of the possibility of discovering true laws based on known facts. The
longer and more despairingly I tried, the more I
came to the conviction that only the discovery of a
universal formal principle could lead us to assured
results. . . . How, then, could such a universal
principle be found?"[5]

This last question was rhetorical and Einstein proceeded to describe how he had intuitively arrived at the special
theory of relativity. By "intuitively" Einstein actually
meant the heuristic methodologies of mathematics:

Nature is the realization of the simplest conceivable mathematical ideas. I am convinced that
we can discover by means of purely mathematical
constructions, those concepts and those lawful connections between them which furnish the key to the
understanding of natural phenomena. Experience may
suggest the appropriate mathematical concepts, but
they most certainly cannot be deduced from it.
Experience remains, of course, the sole criterion
of physical utility of a mathematical construction. But the creative principle resides in mathematics. In a certain sense, therefore, I hold it
true that pure thought can grasp reality, as the
ancients dreamed.[6]

Machism had erred by making science descriptive but
mathematics provided science with a constructive element to
give form and meaning to those speculative impulses which

Einstein considered to be the essence of scientific genius. And, more importantly, it liberated the scientist from his dependence on observation and experimentation.[7]

Mach had detected Platonistic elements in Einstein's early work. Had he lived just a few more years, he would have felt completely justified in rejecting relativity as too metaphysical, for Einstein's idealism became even more pronounced as his successes bred self-confidence. Nevertheless, both Einstein and Mach were dedicated to emancipating science from the "accidents of the individual standpoint and individual personality."[8] And, as was the case with Mach, Einstein's concern over the anthropomorphism of physics was not necessarily scientific at bottom. Like Mach, he had been burdened as a youth with a sense of self-consciousness which seemed to separate him from the world-at-large, and he attributed his desire to become part of something larger than, and transcendent of, the self to the initial religious stirrings of his youth. However, rather than submerge himself in Mach's world of incoherent sensations, Einstein chose a more spiritual, even mystical path. As Einstein so eloquently put it, "science without religion is lame, religion without science is blind."[9] It was this reciprocal relationship between science and religion which had sustained Kepler and Newton through years of hardship and frustration, and Einstein even asserted that it made scientists the only truly religious souls amidst the skepticism and materialism of the modern age.

Einstein's differences with the quantum physicists arose out of his belief that the universe operated according to certain a priori principles, particularly causality. The probabilistic nature of quantum mechanics was necessary only because certain "hidden parameters" determining the exact position of particles were unknown to the experimenter.[10] The quantum theorists responded that, by even suggesting that such "hidden parameters" existed, Einstein was reintroducing metaphysical elements into physics as well as unnecessarily complicating the mathematical formalism of the theory. They even jokingly admonished Einstein for describing God in anthropomorphic terms with his famous dice-throwing analogy, but he remained adamant. Ironically, Einstein discredited himself when, at the Solvay Conference in 1930, he forgot to apply his own general theory to a thought experiment he had designed to deflate the arguments of the quantum physicists. It must have been a sad, even pathetic scene, as the man who had revolutionized physics refused to accept the implications of his own theories, but there was no room for indeterminancy in Einstein's universe.[11]

Einstein was aware that his God and the "cosmic religion" to which he appealed had little in common with the anthropomorphic deities and beliefs of most religions, including Christian tradition. In fact, he looked to science to purify religion of the "dross of its anthropomorphism." Science confirmed man's faith in the rationality of the universe and, in the process, produced a sense of reverence "toward the grandeur of reason incarnate in existence, and which, in its profoundest depths, is unaccessible to man."[12] The fact that the universe could be both comprehensible to reason yet ultimately unfathomable did not seem to trouble Einstein; this paradox in nature merely revealed "an

intelligence of such superiority that compared with it, all
the systematic thinking and acting of human beings is an
utterly insignificant reflection."[13] This sense of awe or
"wonder" before the ineffable produced the humility necessary
for any true religious faith. Thus, science helped to liberate both religion from its anthropomorphism and the individual from his egocentricism, "so that he may place his powers
freely and gladly in the service of mankind."[14]
     Einstein's belief in this reciprocal relationship
between science and religion helps explain away some of the
misunderstanding concerning the moral and philosophical
implications of relativity. As Lewis Feuer noted, the term
"relativity" was more a symbol of generational rebellion than
an accurate description of Einstein's theories. He suggested, as did Einstein himself, that a more appropriate name
would have been the principle of covariance or invariance.
While relativity may have denied the existence of such classical absolutes as time and space, Einstein had no intention
of relativizing the universe when he argued that there were
no privileged frames of reference; he simply meant that all
laws of nature must be independent of any particular frame of
reference so that "invariant laws valid for all observers can
be stated."[15] In this sense, Einstein realized Mach's dream
of freeing science from its egocentric and anthropomorphic
elements. However, sustained by his quasi-religious, quasi-
mystical faith in the harmony and rationality of the
universe, he then moved beyond Mach toward a higher, more
objective synthesis or absolutism which he believed to be
valid for the spiritual as well as the physical realm. As he
once remarked, "Ethical axioms are found and tested not very
differently from the axioms of science." Thus, any attempt
to justify a relativized ethics on the basis of Einstein's
theories would be incorrect not only for scientific reasons,
but because there

     is no room in this [ethics] for the divinization of a
     nation, of a class, let alone of an individual. Are
     we not all children of our father, as it is said in
     religious language? Indeed, even the divinization of
     humanity, as an abstract totality, would not be in
     the spirit of that ideal. It is only to the
     individual that a soul is given. And the high
     destiny of the individual is to serve rather than to
     rule, or to impose himself in any other way.[16]

It was Einstein's belief that the laws governing nature and
society could be reduced to a set of universal principles
independent of both the individual, and man himself, which
formed the foundation for his moral, and political
philosophy.
     This underlying idealism in Einstein's social philosophy
became even more pronounced after World War II, when his re-
forming zeal assumed a romantic, almost quixotic quality. He
continued to call for a world government which, unlike the
League of Nations, had the power to enforce disarmament and
reallocate resources and technology. However, he began to
despair that the threat to civilization and the human race
posed by the mounting stockpiles of atomic weapons could be
averted through political mechanisms. The problems confront-

ing modern man required a new type of thinking which he described as a kind of religious conversion experience, for the "real problem is in the minds and hearts of man. We will not change the hearts of other men by mechanisms but by changing our hearts and speaking bravely." Such a "change of heart" would make men not only "willing but actually eager to submit ourselves to binding authority necessary for world security."[17]

Needless to say, Einstein was criticized as an idealist, or even worse, as a doddering old man who did not understand the realpolitik of the postwar world. And, to a degree, these criticisms were valid, for Einstein was an idealist, just as Lenin had predicted. His was an idealism not of un-realistic hopes and dreams but of a deep and abiding faith that the universe was rational, predictable, and governed by discoverable laws. Furthermore, these laws were predeter-mined and universal, and therefore, independent of any frame of reference. As he once remarked in an interview:

> Everything is determined, the beginning as well as the end, by forces over which we have no control. It is determined for the insect as well as for the star. Human beings, vegetables, or cosmic dust, we all dance to a mysterious tune, intoned in the dis-tance by an invisible piper.[18]

These a priori assumptions were responsible for his divergence from Machian positivism and his quarrel with quantum mechanics. More importantly, these assumptions were fundamental to relativity theory and his belief that mankind could be brought under a single political and moral umbrella. The sublimity of the universe not only inspired Einstein to attempt to reduce nature to "the smallest number of mutually independent conceptual elements;"[19] it also inspired the humility requisite for man in general to over-come his "egocentric cravings" and live according to what Einstein considered to be the only truly universal moral axiom: "Only a life lived for others is the life worth-while."[20]

Notes

1. Lewis Feuer, Einstein and the Generations of Science (New York: Basic Books, 1974), pp. 37 - 38.
2. V. I. Lenin, Materialism and Empirio-Criticism (New York: International Publishers, 1927), p. 34.
3. Phillip Frank, "Einstein, Mach and Logical Positivism," in Albert Einstein: Philosopher-Scientist, Vol. 1, edited by Paul Arthur Schilpp (New York: Harper Brothers, 1959), p. 272.
4. Albert Einstein, "Autobiographical Notes," in Einstein: Philosopher-Scientist, Vol. 1, p. 21.
5. Einstein, "Autobiographical Notes," p. 53.
6. Einstein, Ideas and Opinions (New York: Crown Publishers, 1954), p. 370; Gerald Holton, "Mach, Einstein, and the Search for Reality," Daedalus 97 (Spring 1968): 650 - 51.
7. Holton, "Mach, Einstein, and the Search for Reality," p. 656.
8. Ernst Cassirer, Substance and Function, and the

Einstein's Theory of Relativity (New York: Dover Books, 1953), p. 445.
9.   Einstein, "Science and Religion," in *Ideas and Opinions*, p. 46.
10.   Bernard d'Espagnat, "The Quantum Theory and Reality," *Scientific American* 241 (November 1979): 165.
11.   Ronald W. Clark, *Einstein: The Life and Times* (New York: World Publishing Co., 1971), pp. 337 - 47; Neils Bohr, "Discussion with Einstein on Epistemological Problems in Modern Physics," in *Albert Einstein: Philosopher-Scientist*, Vol. 2, pp. 201 - 41.
12.   Einstein, "Science and Religion," p. 49.
13.   Einstein, "The Religious Spirit of Science," in *Ideas and Opinions*, p. 40.
14.   Einstein, "Science and Religion," p. 42.
15.   Feuer, *Einstein and the Generations of Science*, p. 61; Elie Zahar, "Mach, Einstein, and the Rise of Modern Science," *British Journal of Philosophy of Science* 28 (Summer 1977): 195 - 213.
16.   Einstein, "Science and Religion," p. 43.
17.   *New York Times*, June 23, 1945, Sect. 6, p. 7.
18.   Clark, *Einstein: Life and Times*, pp. 346 - 47.
19.   Einstein, "Religion and Science Irreconcilable?", in *Ideas and Opinions*, pp. 48 - 49.
20.   Virgil Hinshaw, "Einstein's Social Philosophy," in *Albert Einstein: Philosopher-Scientist*, Vol. 2, p. 650.

4.

Einstein and Epistemology

BURTON H. VOORHEES and JOSEPH R. ROYCE

In a conversation recorded by A. Moskowski (1970)
Einstein responded to a question concerning the reasons for
his scientific success with the statement: "I am as stubborn
as a mule." He paused for a moment, then continued: "I am
stubborn, and I have a good nose." In his "Autobiographical
Notes" (1949) he repeats his olfactory metaphor. Giving
reasons for his choice of physics over mathematics as a
specialization he states: "In [physics] I soon learned to
scent out that which was able to lead to fundamentals and to
turn aside from everything else, from the multitude of things
that clutter up the mind and divert it from the essential."
What was the nature of this sense of smell which directed Einstein's seeking? Was it a unique gift, a freak of
heredity, or is it something that we all possess, perhaps not
in the degree of an Einstein, but as an ability or talent
that can be trained by appropriate techniques? If this
latter is the case, then it is important to seek a rational
description of this "sense for the essential" as a first step
to developing training methods. In this chapter we will discuss the nature of this intuitive sense from the viewpoint of
psychological epistemology.
Einstein was not a professional philosopher and felt no
need to leave a detailed statement of his epistemic methods.
In fact, he held that the working scientist could not afford
to maintain too strong a commitment to any epistemic system.

Epistemology without contact with science becomes an
empty scheme. Science without epistemology is . . .
primitive and muddled. However, no sooner has the
epistemologist, who is seeking a clear system, fought
his way through to such a system, than he is inclined
to interpret the thought-content of science in the
sense of his system, and to reject whatever does not
fit. The scientist cannot afford to carry his
striving for epistemological systematic that far
external conditions, which are set for him by the
facts of experience, do not permit him to let himself
be too much restricted by the adherence to an
epistemo-logical system ("Autobiographical Notes,"
1949).

Thus it seems more natural to speak of Einstein's epistemic attitude than of his position. We recognize, of course, that Einstein's attitudes changed over the course of his life. He did not study mathematics with any enthusiasm as a student at the Zurich Polytechnique because of a positivistic conviction that physics did not need abstract terms. When he realized the relevance of the tensor calculus for his work towards the general theory of relativity, he abandoned this belief. In this respect his "Autobiographical Notes," his self-proclaimed 'obituary,' must be viewed as containing his final epistemic statements, made with the benefit of hindsight.

Despite his direct disclaimer of epistemic fidelity, various philosophical camps have vied to claim Einstein as one of their own. Philipp Frank (1949), for example, states that both metaphysicians and positivists have claimed Einstein as their epistemological exponent. Hans Reichenbach (1949), in a paper praised by Einstein, brushes aside Einstein's personal statement to Reichenbach that his achievements were based on his "profound belief in the Unity of nature" to claim that he (Einstein) was a radical empiricist. Neo-Kantians, on the other hand, can trace their origins in part to Einstein's principle of general covariance which, as masterfully expounded by Ernst Cassirer (1953), requires revision of the Kantian categories of space and time. The truth of the matter is that Einstein was a self-confessed epistemic opportunist, borrowing anything that would suit his purpose. He has acknowledged the influence of Mach, Kant, and many others on his thinking.

> I see on the one side the totality of sense-experiences, and, on the other, the totality of the concepts and propositions.The relations of the concepts and propositions among themselves and each other are of a logical nature, and the business of logical thinking is strictly limited to the achievement of the connection between concepts and propositions among each other according to firmly laid down rules, which are the concern of logic. The concepts and propositions get "meaning," viz. "content," only through their connection with sense experiences. The connection of the latter with the former is purely intuitive, not itself of a logical nature. The degree of certainty with which this connection, viz., intuitive combination, can be undertaken, and nothing else, differentiates empty fantasy from scientific truth ("Autobiographical Notes," 1949, p. 11 -13).

Thus, Einstein is neither empiricist, rationalist, nor rational-empiricist. Nor is he any mere mystical intuitionist, for he strongly insists on the necessity of both rational and empirical validation of truth claims. His final court of appeal, however, is intuition. Logic and experience play an indispensable role of support, but intuition is the cutting edge. The role of logic and experience is to help in selecting those intuitive combinations of precept and concept possessing the greatest degree of certainty.

Further, Einstein's statements are not particularly

idiosyncratic. They characterize a particular attitude found
among many eminent theoretical physicists. Despite any lip
service paid to logical positivism, this attitude is in
essence metaphoric. Banesh Hoffmann, a former research associate of Einstein, and a prominent relativist in his own
right, states:

> The essence of Einstein's profundity lay in his
> simplicity; and the essence of his science lay in
> his artistry — his phenomenal sense of beauty
> (1972, p. 3).

Henry Margenau, physicist and philosopher of science, writes:

> [In] the actual practice of physicists . . .
> deductive uncertainly is held in bounds by
> important habits of reasoning, by pre-empirical
> commitments to certain forms of theory — in short,
> by factors not imposed and frequently not even
> suggested by the facts themselves. Philosophers
> have spoken of them as categories of thought, as
> razors that shear away irrelevancies of explana-
> tion, and as injunctions enforced by a lumen
> naturale or by divine revelation; physicists have
> used phrases like economy, . . . simplicity, and
> elegance (1960, p. 54 - 55).

The well-known Indian physicist, E. G. C. Sudarshan,
writes: We recognize creativity, the seeing of things as
they really are, the flash of lightning that illuminates the
lay of the land, the incredible that with the flux of time
appears as the inevitable." And Paul Adriene Maurice Dirac,
who has been called the greatest British physicist since
Newton, goes so far as to say, "It is more important to have
beauty in one's equations than to have them fit experiment"
(1963, 45).   These and many other statements indicate the
importance of metaphoric considerations to individual
scientists at the top of their profession. Terms such as
elegance, simplicity, economy, and beauty are equally
applicable to scientific theories, mathematical equations,
and works of art. For these to have validity in support of a
metaphoric knowledge claim, however, something more is
required.   Beauty, we are assured, lies in the eye of the
beholder. Thus, if the metaphoric criterion of universality
is to be satisfied, there must be agreement among a community
of informed beholders that a particular thing is indeed,
elegant, simple, economical, and beautiful. The metaphoric
criterion is based on informed agreement as to what is
aesthetically pleasing, just as rational and empirical
criteria are based on informed agreement as to what
constitutes a rational statement, or an accurate observation.
To capture the imagination of a scientific community a
theory must possess an intrinsic intuitive appeal. It is on
this essentially aesthetic point of intuitive obviousness
that, according to Thomas Kuhn, scientific revolutions turn.
And once a paradigm is established there is a basis for
induction. The paradigm specifies the kind of patterns to be

expected in the data, and knowing what to look for gives a
guide for discovery. The rational-empirical process of
Figure 1 is sufficient to conceptualize the advance of
paradigmatic science. Metaphoric considerations gain
importance in times of paradigm change. Thus, we are led to
postulate the cyclic psycho-epistemic process of Figure 1 as
giving a more complete description of how science
progresses. This figure indicates that hypotheses are not
induced directly from data, they are induced only in the
context of a given paradigm. The hypotheses are then used,
as in the rational-empirical cycle, to generate predictions
that can be empirically tested. Confirmation or refutation
of predictions is taken as supporting or falsifying the
induced hypotheses rather than the accepted paradigm. <u>Only
when the hypothesized theoretical structure necessary to
generate predictions that match experiment becomes
excessively obtuse and cumbersome does the potential for
emergence of a new paradigm arise</u>. The paradigm provides a
lens, as it were, to focus intuition. It gives a set of
criteria for what forms of theory are to be considered
elegant, simple, economical, and beautiful. But its ultimate
validity is metaphoric rather than rational or empirical.
The advantage of conceptualizing scientific knowledge
acquisition in terms of Figure 1 is that the necessary
existence of a paradigm is explicitly acknowledged and thus
the occurrence of a paradigm shift becomes rationally
comprehensible rather than accidental or unpredictable.

So far we have described the epistemic process as it
might become manifest in a scientific community. The other
half of psychological epistemology involves the internal
psychological processes that occur within each individual
knower. Because of Einstein's emphasis on the need for
intuition in order to solve the problem of induction, at
least in cases in which there is no accepted paradigm to
guide the way, it is this which we shall make the focus of
our discussion. We realize, of course, that we cannot hope
to give an adequate answer to the question of what intuition

Figure 1

The Psycho-Epistemic Cycle

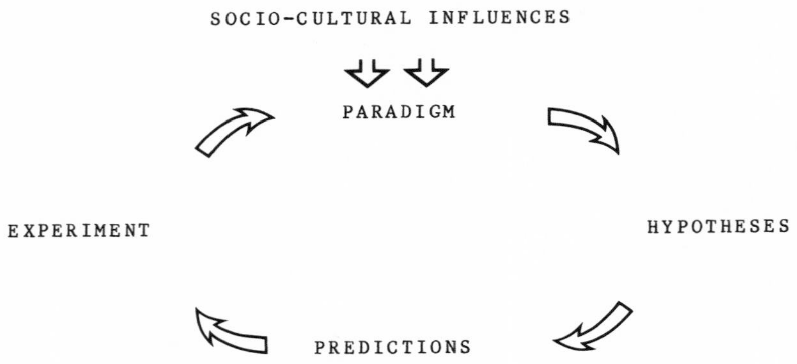

SOCIO-CULTURAL INFLUENCES

PARADIGM

EXPERIMENT

HYPOTHESES

PREDICTIONS

is and how it operates. Nevertheless, it may be possible to point out some directions. As Einstein indicated, one of intuition's functions is to make the connection between precepts and concepts. It is generally acknowledged that Einstein's rational and empirical skills were those of a highly competent professional. But his genius, as indicated in the earlier quote from Hoffman, lay in his remarkable intuition. Elsewhere Hoffmann comments:

> When we see how shaky were the ostensible foundations on which Einstein built his theory, we can only marvel at the intuition that guided him to his masterpiece. Such intuition is the essence of genius. Were not the foundations of Newton's theory also shaky? . . . And did not Maxwell build on a wild mechanical model that he himself found unbelievable? By a sort of divination genius knows from the start in a nebulous way the goal toward which it must strive. In the painful journey through uncharted country it bolsters its confidence by plausible arguments that serve a Freudian [sic] rather than a logical purpose. These arguments do not have to be sound so long as they serve the irrational, clairvoyant, subconscious drive that is really in command (1972, p. 127-128).

We distinguish between ingenuity and intuition. The former serves the puzzle-solving activities of paradigm or normal science; it enables the scientist to generate ingenious hypotheses. The latter, as Hoffmann states, guides the revolutionary scientist in his journey into the unknown. The epistemic process we have presented is a process that is manifest within a community of scientists — only rarely does any individual scientist contribute to all stages of this process. If a paradigm is given, this process operates to elaborate the more detailed implications.In such cases of normal scientific advance it is ingenuity, skill at combining aspects of the given, which predominates. Intuition — as recognition of what was previously unknown — is more important in times of paradigm change. Analysis of the thought processes that guided Einstein's search for special and general relativity provides insight into the nature of his intuition. As reported by M. Wertheimer (1959), Einstein's scientific progress was oriented by a focus on certain thought experiments (Gedankenexperiment) that served the function of providing a standard against which tentative conclusions could be tested. Special relativity, for example, was catalyzed by the question: would a person traveling at the speed of light see a light ray as a standing wave? As it happened, this question asked by a sixteen year old, initially perhaps in a moment of idle curiosity, was of deep physical significance. It cannot, however, be considered a standard question of normal science. It is not well posed, that is, it is not a puzzle which a scientist is assured of being able to solve given enough ingenuity. Rather, it asked fundamental questions about the relationship of mechanics and electrodynamics and did so in a sufficiently vague form that the answer, if it existed, was not automatically restricted-in-the-asking to

lie within the then existing paradigms of mechanics or electrodynamics. It thus provided a center around which Einstein
could indulge the free conceptual play in which he
delighted — without prejudice as to expected outcome. After
much effort and several blind alleys Einstein was driven,
although for consistency we might better say drawn, to a recognition of two basic principles: the equivalence of all
inertial frames, and the constant speed of light. Taking
these as fundamental led to the rapid formalization of
special relativity theory. In a similar way his famous
elevator _Gedankenexperiment_ led to the principle of equivalence, and provided a catalyst in construction of the
general theory of relativity. We may ask, however, where in
this process is intuition? Is it in the asking of profound
questions, in the gestalt recognition of the conceptual
implications inherent in postulated general principles
catalyzed by consideration of those questions, or in the act
of recognizing those principles?

A review of scientific and philosophic positions on
intuition has been given by M. R. Westcott (1968), while a
critical attack on the appeal to intuition as a means of
validating knowledge claims is made by M. Bunge (1962).
Westcott distinguishes a variety of positions ranging from
what he calls classical intuitionism, exemplified by Spinoza,
Bergson and Croce, to an almost complete denial of the existence of intuition, as is the case with Bunge. The classical
intuitionists are characterized as believing that intuition
is the only possible way to gain certain knowledge of truth,
beauty, and ultimate reality. More moderate modern intuitionists hold that intuitions may be incorrect and should be
checked by other than intuitive means. Bunge argues that
intuition is nothing more than change association, something
that scientists may, if they are inclined, indulge privately
but should never discuss in polite scientific company.

Bunge polemicizes against philosophical intuitionism,
claiming it is sterile and supports dogmatism and authori-
tarianism. He is not, however, able to eliminate intuition
from scientific thinking, and so attempts to do the next best
thing, which is to classify it. He provides a list of ten
different forms in which intuition may become manifest in
scientific thinking. If we combine his categories of common-
sense and good judgment, the nine resulting categories can be
classified according to the cognitive processes in Royce's
three ways of knowing scheme, as in Table 1. Thus, for
example, what Bunge calls quick identification is for us the
perceptual aspect of empirical intuition. Similarly, we
identify Bunge's skill in forming metaphors with the conceptual aspect of metaphoric intuition, and Bunge's power of
synthesis with the symbolic aspect of rational intuition.

Reading this table vertically tells us, for example,
that it may be useful to view empirical intuition as involving, in a first approximation, cognitive processes associated
with "quick identification" (of forms and characteristics, a
perceptual process), a "clear understanding" (the grasp of
relationships in identified sensations, a conceptual
process), and "interpretation ability" (assignment of meaning, a symbolic process). Reading horizontally suggests
viewing the perceptual aspect of intuition, for example, as
involving "quick identification," "common sense" (as

Table 1

Aspects of Scientific Intuition

	Empirical Intuition	Rational Intuition	Metaphoric Intuition
Perceptual Aspect	Quick Identification	Common Sense, Good Judgment	Representational Ability
Conceptual Aspect	Clear Understanding	Catalytic Inference	Skill in Forming Metaphors
Symbolic Aspect	Interpretation Ability	Power of Synthesis	Creative Imagination

abstracted from experiences stored in memory), and "representational ability."

Taking a different approach, we can trace the origins of the word intuition through the old English intuycion (contemplation), the old French intuition(view, contemplation), the Latin intueri (to look at or toward, contemplate), and finally to the hypothesized proto-Indo-European form teu (to pay attention to, turn to). This, taken together with Polanyi's (1958-1966) concept of tacit knowledge and his likening of tacit knowing to the gestalt phenomena of closure, suggests that a visual metaphor may be appropriate. For us intuition is not in itself a thing or an ability. It is an awareness of significance, or of potential significance. Following one's intuition simply means moving in directions that provide an increased sense of meaning. An intuition occurs when an array of seemingly unrelated objects of consciousness are seen to form a unified and meaningful whole. If we follow Einstein and regard scientific intuition as providing the bridge that links precepts and concepts, then what is recognized is that some class of perceptions exemplifies a relatively closed system of conceptual relationships. It is the closure of the conceptual system that provides the sense of unity, while meaning arises through recognition of patterns of relationship between the newly perceived whole and other known objects of consciousness. Pushing our metaphor further, we can say that the quality of an intuition, which must be judged rationally and empirically, will depend on clarity of perception; or, taking the most generalized sense of perception, on clarity of consciousness. In speaking of the acuity of Einstein's intuition we are speaking, metaphorically, of the clarity of his vision.

These heuristic statements can be given substance by reference to table 1. That is, we posit that the entries in this table represent ongoing psychological processes which lay the foundations for the occurrence of an intuitive event. This event, following Polanyi, may be likened to the phenomenon of emergence of a gestalt figure from a perceptual ground. From this it may be possible to draw clues as to possible methods for training more accurate intuitions.

We do not, however, as Einstein most certainly did not, claim that intuition provides certain knowledge of reality. Nor are we saying that metaphoric truth claims can only be validated intuitively. The relation of intuition to psychological epistemology is well illustrated in the tale of the Incomparable Celestial Apple, related by Idries Shah (1971):

A zen student approached his teacher and requested to be shown a real object of the Celestial realm. Perhaps one of the Apples of Paradise. The teacher snatched an apple from a bowl of fruit beside him and held it under the student's nose.

The student objected that this particular apple could not be of celestial origin — it was partially rotten.

"Yes," replied the master, "Living as we do in this world of decay, this is as close as you will ever get to the Apples of Paradise." And so the need for the paring knife of reason and experience.

Suppose that our student in this tale had never seen an apple before. Perhaps he had some concept of apple, but now he must recognize this thing held under his nose as an actual apple. That, according to Einstein, requires intuition. Having achieved this recognition his reaction will depend on what beliefs he has about apples, and these beliefs are judged in part by the metaphoric criterion of universality. Perhaps his belief is that apples are good to eat. Then he must cut away the rotten part. This requires discrimination, that is, reason. And then he must perform the experiment of eating. Each of these three ways of knowing gives a different kind of information.

But note that the entire process requires intuition. The object must first be recognized. And, as Einstein has said, the fact that such recognition is possible is the deepest mystery of nature.

Bibliography

Bunge M. Intuition and Science, Englewood Cliffs, N.J.: Prentice-Hall, 1962.
Cassirer, E. Substance and Function and Einstein's Theory of Relativity, Translated by William Curtis Swabey and Marie Collins Swabey. (New York: Dover, 1953.
Darwin, C. The Autobiography of Charles Darwin, Edited by Francis Darwin. New York: Dover, 1958, pp. 53 - 54.
Dirac, P. A. M. "The Evolution of the Physicist's Picture of Nature" Scientific American (May 1963) p. 45.
Einstein, A. "Autobiographical Notes" In Albert Einstein: Philosopher-Scientist, Vol. 1 (edited by P. A. Schilpp.) New York: Harper and Row, 1949.
Feyerabend, P. Against Method: Outline of an Anarchistic Theory of Knowledge London: Verso, 1978.

Frank, P. "Einstein, Mach, and Logical Positivism" In
Albert Einstein: Philosopher-Scientist, Vol. 1. edited by
P. A. Schillp. (New York: Harper and Row, 1949).

Hoffman, B. Albert Einstein: Creator and Rebel. New York:
Viking Press, 1972.

Kuhn, T. The Structure of Scientific Revolutions. 2d ed.
enl. Chicago: University of Chicago Press, 1970.

Margenau, H. "Does Physical 'Knowledge' Require A Priori or
Undemonstrable Presuppositions?" In The Nature of Physical
Knowledge, edited by L. W. Friedrich and S. J. Friedrich.
Indiana University Press: Bloomington, 1960.

Moskowski, A. Conversations with Einstein Translated by
H. L. Finch. London: Sidgwick & Jackson, 1970.

Polanyi, M. Personal Knowledge: Towards a Post-Critical
Philosophy. Chicago: University of Chicago Press, 1958.

Polanyi, M. The Tacit Dimension. Garden City, N.Y.:
Doubleday, 1966.

Reichenbach, H. "The Philosophical Significance of the
Theory of Relativity" In Albert Einstein:
Philosopher-scientist Vol. I, edited by P. A. Schilpp.
New York: Harper and Row, 1949.

Royce, J. R. The Encapsulated Man. Princeton, N.J.: Van
Nostrand, 1964.

Royce, J. R. "Cognition and Knowledge: Psychological
Epistemology" In Historical and Philosophical Roots to
Perception, edited by E. C. Carterette and M. P.
Friedman. New York: Academic Press, 1974, pp. 149 - 176.

Royce, J. R. "Three Ways of Knowing and the Scientific
World-view "Methodology and Science 11 (1978): 146 - 164.

Royce, J. R., H. Coward, E. Egan, F. Kessel, and L. Mos
"Psychological Epistemology: A Critical Review of the
Empirical Literature and the Theoretical Issues," Genetic
Psychology Monographs 97 (1978): 265 - 353.

Shah, I. The Sufis. Garden City, N.Y.: Doubleday,
1971.

Sudarshan, E. G. C. "The Temper of Science" in Science and
Technology in India, edited by B. R. Nanda. New Delhi:
Vikas Press.

Westcott, M. R. Toward a Contemporary Psychology of
Intuition. New York: Holt, Rinehart, and Winston, 1968.

Wertheimer, M. Productive Thinking. (New York: Harper,
1959.

Part III

Religion

5.
Religion, Relativity, and Common Sense: Einstein and the Religious Imagination
WILLIAM F. LAWHEAD

To fully understand Einstein's contribution to the
thought and life of the twentieth century, we cannot under-
estimate the impact he had upon the religious imagination.
The word "imagination" is key here, for it was in the imagination that Einstein had his greatest point of contact with
the average layperson. Though the general public could not
understand Einstein's equations, they sensed that in these
symbols there was to be found the picture of a strange and
bewildering universe. The thought of this provoked both
hostility and awe.

In spite of the numerous popularizations of Einstein's
theory, it was the form or style of the theory rather than
its content that was the source of most of the religious responses, whether they were negative or positive. Einstein's
theory, when read between the lines, seemed to provide scientific confirmation for either the deepest of human fears or
the highest of human hopes. For example, its radical viola-
tion of traditional conceptions was said to imply either a
rationally lawless universe on the one hand, or a comforting
mystical one on the other. Similarly, Einstein's style of
doing science was identified as both impious, vain speculation and as the exercise of a bold, imaginative faith. In
what follows, I hope to make clear the elements in the theory
of relativity that led to both hostile and sympathetic reactions.

Many religious perspectives at the turn of the century
carried a heavy commitment to common sense. This can be
measured by studying the widespread acceptance and promul-
gation of Scottish common sense realism in America. It is
said to have served as "the handmaiden of both Unitarianism
and Orthodoxy.[1] The entanglement of this common-sense
epistemology with theological method is clearly illustrated
in the works of Charles Hodge who taught theology at
Princeton for fifty years. Over 3,000 of his students
carried the philosophy and theology of common sense to
churches, seminaries, academies, and colleges all over
America.[2]

In his influential Systematic Theology, published in the
late nineteenth century, he linked biblical theology with a
Baconian view of science. In one of the opening sections of
this work he states, "The Bible is to the theologian what

nature is to the man of science. It is his storehouse of
facts. . . . In theology as in natural science, principles
are derived from facts, and not impressed upon them."[3]
Einstein's disregard for this commonly accepted methodological principle would prove to be a major cause of antago-
nism toward the new theory for many experimental scientists
and religionists.

As Hodge put it: "Confidence in the well-authenticated
testimony of our senses, is one of those laws of belief which
God has impressed upon our nature. . . . Confidence in our
senses is, therefore, one form of confidence in God."[4]

However, if Einstein was right, the bulk of our ordinary
experiences were deceptive and our ordinary ways of thinking
were hopelessly out of touch with the essential scheme of
things. The Psalmist had said, "The heavens declare the
glory of God...," but the heavens of the poet's perceptions
and those of Einstein's calculations seemed to be totally
different worlds. There had been problems enough four
centuries ago when Copernicus had exchanged our God-given,
earth-bound perspective for one of more scientific convenience. Now, an upstart scientist was asking us to accept a
perspective on things that went beyond anything that can be
seen or imagined. Not only our normal point-of-reference
within space, but our whole experience of time and space had
to be abandoned.[5]

This contrast between the world as it is given to us in
our concrete, ordinary experience and the world flowing from
the abstractions of relativity made otherwise innocent scientific articles take on the appearance of antireligious
tracts. This dichotomy was, no doubt, the basis for Cardinal
O'Connell's oft-quoted charge in 1929 that Einstein's view of
space and time is "a cloak beneath which lies the ghastly
apparition of atheism." It is significant that this judgment
was made before the appearance of Einstein's widely read
articles on religion in American periodicals more than a year
later. Thus, the Cardinal drew his harsh conclusion from his
perception of the implications of the scientific theory itself. He saw the complexities of Einstein's theory as
leading students away "into a realm of speculative thought,
the sole basis of which . . . is to produce a universal doubt
about God and His creation."[6]

For centuries, natural theology had attempted to base a
secure moral and religious order upon the rational foundations of the physical order. If time and space were not
absolutes, how could we hope to find ethical and religious
absolutes? Thus, for many, theological notions had become so
solidly married to the Newtonian worldview that religion was
now threatened with the loss of its metaphysical foundations. In an interview published in the Saturday Evening
Post, Einstein repudiated the philosophical sense of "relativity" view, but the writer still went on to say:
"Einstein stands in a symbolic relation to our age — an age
an age characterized by a revolt against the absolute in
every sphere of science and of thought."[7]

Not all of the responses during the 1920s were this
negative, however. One of the very first religious responses
appeared in a Protestant journal in 1922. After explaining
the new theory, the author jumped from the finiteness of the
universe to conclude its dependence upon a spiritual

absolute.   In his conclusion, he confidently assured his
readers that "a finite and temporal, and therefore <u>created</u>,
universe issues also from this latest scientific-philosophic
world-view."[8]
     Some clergy found confirmations of even more specific
religious doctrines in the new theory.   The Reverend Henry
Howard of the Fifth Avenue Presbyterian Church in New York
said in a sermon that "the newest hypothesis of the German
mathematician was proof of what was guessed at in St. Paul's
synthesis that all things were one."[9]   Similarly, another
speaker claimed that Einstein was close to agreement with the
Christian Science teaching that "all is infinite mind and its
infinite manifestation...there is no matter."[10]   In a response to Cardinal O'Connell's attack on the scientist, Rabbi
Goldstein said that Einstein's discovery of the ultimate
unity of things provides a "scientific formula for mono-
theism."[11]   He suggested that the Cardinal's trinitarianism
and not belief in God may have been the real issue at stake.
     The years 1929 and 1930 mark a significant juncture with
respect to the religious impact of Einstein's thought.
During this period the popular press presented a flurry of
widely publicized comments and articles from Einstein stating
his personal, religious views.   For example, in April 1929,
Einstein made the now familiar statement to Rabbi Goldstein
as a reply to the charges of atheism made against the scientist: "I believe in Spinoza's God who reveals himself in the
harmony of all that exists, not in a God who concerns himself
with the fate and actions of men."[12]   Most significantly, the
November 9, 1930 <u>New York Times Magazine</u> contained a lengthy
article by Einstein entitled "Religion and Science."   In
this, he spoke movingly of the "cosmic religious feeling" and
its importance for both science and life in general.   On the
other hand, however, he unequivocally rejected the notion of
a personal God and most of its attendant conceptions found in
the Judeo-Christian tradition.
     Typical of the confused responses of this period were a
number of critical articles and editorials against Einstein
which were published in <u>Commonweal</u>.   With their conservative
Catholic perspective, its editors reacted strongly to
Einstein's rejection of a personal God.   However, their
attacks seemed to fault indiscriminately both his theology
and his physics as though no real differentiation needed to
be made.   One of their first editorials of this sort stated
that "Einstein's fame is one of the most paradoxical phases
of modern publicity.   It illustrates once more how blind
faith and mere credulity, not to mention superstition, still
guide and control the mental processes of so many people."[13]
The confusion appears in the fact that while it is clearly
his view of God as expressed in the "What I Believe" article
that they are rejecting, they still found it relevant to discredit his scientific fame.
     While the criticisms from conservatives took on a new
intensity, those who had a more pliable theology were delighted by the reaffirmation of the religious spirit from
within the ranks of science itself.   Beleaguered believers
had for too long felt that they were losing the battle to the
infidel scientists, such as T. H. Huxley or Ernst Haeckel,
who claimed that the quest for scientific truth required the
abandonment of religious faith.   There was now the reassuring

example of Einstein who showed that scientific inquiry could not only coexist with religious convictions but that the former grew out of the latter. Typical of Einstein's comments on this is the following: "I am of the opinion that all the finer speculations in the realm of science spring from a deep religious feeling, and that without such feeling they would not be fruitful."[14] A decade later he summed up his position in a memorable phrase: "Science without religion is lame, religion without science is blind."[15] Again, in an article in Scientific American, he claimed that "every true theorist is a tamed metaphysicist" because he necessarily has the conviction that "the totality of all sensory experience can be 'comprehended' on the basis of a conceptual system built on premises of great simplicity."[16] Einstein went on to admit unashamedly that this was a "miracle creed."

Not only does science flow from a deep wellspring of faith and religious sentiment, but the process of science, according to Einstein, relies upon intuition and the creative imagination.[17] He instructed the physicist "to give free reign to his fancy, for there is no other way to the goal."[18] A continual theme of his was that the basic axioms of physics could not be derived from any experimental results but had to be "freely invented."[19] Based upon Einstein's testimony, science no longer seemed to be a plodding, impersonal accumulation of dead, dull facts. Instead, we are told, the mind of the scientist is "akin to that of the religious worshipper."[20] Faith, hope, conviction, passion, imagination, and intuition, long thought by scientific materialists to be the pathetic props of religious belief, seemed now to be vindicated by virtue of the important role they played in Einstein's account of scientific activity. While this was disturbing to the sober-minded experimentalists within the scientific community, the religious public saw something no longer alien to a religious stance toward the world. Though not referring specifically to Einstein, the mood of many was captured in a newspaper report of a speech by Pope Pius in which he said that "harmony between religion and science was 'ever more luminously' confirmed by each new scientific conquest."[21] While the difficulty of understanding Einstein was an initial barrier to his acceptance, this difficulty soon became, for many a proponent of religion, an appealing feature of his theory. As Einstein became established as an unparalleled authority in science, a certain mystique was built up around him. The implications of this for religion were two-fold. First, authority as a source of knowledge has long been fundamental to many religious perspectives. Some truths have to be accepted and believed on faith. Indirectly (and against the tenor of his own beliefs), Einstein lent plausibility to this view. Second, Einstein's authority and expertise in mathematics tended to invest his opinions with a similar authority in religion and other areas as well. Interviewers continually badgered him for opinions and advice on all kinds of unlikely topics.

Einstein's reception as a modern-day prophet is indicated not only by those who revered him, but also by those who were horrified by this development. In a 1933 editorial, The Nation printed the following lament in reaction to Einstein's semipopular Herbert Spencer lecture:

We realize rather uncomfortably that Einstein's attitude does tend once more to create a sort of priestly class. Formerly the scientist was merely a man who had a great deal of knowledge of the same sort as that of which we all had a little. He used the instruments which we used, even though he used them more skillfully, and he was as far as the lay-man from any comprehension of the ultimate mystery. Today, the mathematician belongs to a special class of illuminati. We see through the dark glass of our purely human understanding, but he sees face to face. God is an equation.[22]

Though Einstein's cultural impact cannot be divorced from the qualities of the man himself, it is important to realize that Einstein's ideas could not be communicated to the public through a straightforward reporting of the facts as with most of the familiar scientific discoveries made prior to the twentieth century. The indirect methods of analogies, allegories, metaphors, parables, and stories have always been necessary to communicate the difficult and tran-scendent truths of religion. Now, quite clearly, the physicists were forced to resort to these devices. Their alternatives were limited and uncomfortable. On the one hand, they could communicate in the language of mathematics. Though having the virtue of being precise and accurate, this did not help to anchor the new theory in the world in which the ordinary person lived. On the other hand, physicists could attempt to use an analogy or illustration to make their point. But this had its drawbacks also, for any analogy used was necessarily bathed with numerous qualifications and the disclaimer added that the example only gave a partial glimpse into what they were really trying to say. Many a physicist, no doubt, could feel empathy with the Apostle Paul who con-fessed, "What we see now is like a dim image in a mirror.... What I know now is only partial."[23]

To the bewildered layperson, it seemed as though reason and language could not contain the truths about physics any more than they could the mysteries of religion. The result of all this was that the bounds of the believable became con-siderably expanded and the requirements for intelligibility became considerably diminished in the mind of the public. Consequently, in a number of religious articles, the paradoxes of relativity were compared with such traditional though difficult Christian truths as the Trinity and immor-tality. The object was to show that the Christian doctrines compared favorably with the new scientific ones. For example, we find the following argument in the Hibbert Journal in 1940:

These Alice-in-Wonderland attributes of space-time merely indicate that our reason cannot deal with a four dimensional continuum. It is based on the three dimensional experience of this world, and cannot go beyond it. . . .
    On this showing a description of the Deity as being One in Three and Three in One is not to be rejected, merely because it is formally illogical. It relates to a state or continuum in which neither

number nor logic holds a meaning.
    It is curious to reflect that scientists, who
had no other motive than the elucidation of a
puzzle of nature, should have stumbled on a truth
which is as essential to the interpretation of
religion as it is to the interpretation of the
physical world.[24]

More recently, a Christian apologist has appealed to the
alleged "looseness" of the Einsteinian universe in a defense
of miracles.

But can the modern man accept a "miracle" such as
the Resurrection? The answer is a surprising one:
The Resurrection has to be accepted by us just
because we are modern men — living in the
Einsteinian-relativistic age. For us, unlike
people of the Newtonian epoch, the universe is no
longer a tight, safe, predictable playing field in
which we know all the rules. Since Einstein, no
modern has had the right to rule out the possibility of events because of prior knowledge of
"natural law." The only way we can know whether an
event can occur is to see whether in fact it has
occurred.[25]

Of course, this account ignores Einstein's own belief in
universal causation and a closed universe of natural
events.[26] Nevertheless, it does demonstrate that the theory
of relativity has become, in the popular mind, more than a
piece of physics telling us about the nature of the world.
Over and above this, it has taken on the stature of a symbol
of a science which is supportive of our search for the ultimate meaning of the world.
    Perhaps the greatest contribution of Einstein's theory
to religious thought lies in the more sophisticated understanding of the nature of scientific knowledge it made
possible. Typically, the popular press tended to personify
science into a self-assured, singular authority with such
phrases as "Science says. . . ." To the public, therefore,
it was remarkable to see noted physicists engaged in heated
debates over the fundamental principles governing the
physical world. As Bertrand Russell described the situation,
"The new philosophy of physics is humble and stammering where
the old philosophy was proud and dictatorial."[27] The lack of
finality and certainty appearing in a science as rigorous as
physics was instructive and brought it a bit closer to the
rest of our distinctively human engagements in the eyes of
the public. Furthermore, Einstein's frequent assertion of
the tremendous distance between the directly observable and
the conceptual systems necessary to understand and interpret
it suggested that theology may be more analogous to science
than we could have ever imagined before.[28]
    To conclude, we should say that it is always difficult
to accurately ascertain what was discussed among the general
populace as they gathered together in barbershops or drawing
rooms. Nevertheless, we can assume that the content of
sermons, editorials, and popular articles had a formative
effect upon the public mind as much as it reflected the con-

cerns and issues of the day. We have tried to suggest that
what the public was responding to was primarily the style,
the mood, or the philosophical flavor suggested by the theory
of relativity. The average layperson was not overly concerned with the question of whether or not there is an ether
or whether gravity and inertia are the same thing.   How
Einstein did science, the spirit animating his research, and
the assumptions underlying it, were more important to the
public than his theoretical discoveries as such.
    There had long been a growing apprehension that imper-
sonal, scientific reason would, to quote Keats,

> Conquer all mysteries by rule and line,
> Empty the haunted air, and gnomed mine —
> Unweave a rainbow. . . .[29]

However, in the aftermath of Einstein's discoveries, the New
York Times editor could proclaim that the scientist and the
poet now complemented one another.[30] Einstein himself firmly
believed that the theory of relativity had no philosophical
or religious bearing that would make any difference to the
average layperson.[31] Nevertheless, the dreams and the sensibility of this century have not remained unaffected.
    The Harper's editor spoke fittingly when he complained
in 1929, "We see an extraordinarily animated public interest
in an alleged discovery which hardly anyone understands"[32]
If nothing else, the astonishing revolution in scientific
thought brought about by Einstein serves to remind us that
the universe we live in can still be full of surprises and
that the best of human knowledge is still all too human.
This awareness alone is a soil sufficiently rich for the
religious imagination to take root and to flourish in it.

Notes

    1.   Sydney E. Ahlstrom, "The Scottish Philosophy and
American Theology," Church History 24 (1955): 257.   Cf.
Theodore D. Bozeman, Protestants in an Age of Science (Chapel
Hill: University of North Carolina Press, 1977).
    2.   See Sydney E. Ahlstrom, "Theology in America: A
Historical Survey," In The Shaping of American Religion,
eds. James W. Smith and A. Leland Jamison, (Princeton:
Princeton University Press, 1961), pp. 232 - 321.
    3.   Charles Hodge, Systematic Theology, 3 vols. (London:
Thomas Nelson and Sons, 1978; New York: Charles Scribner &
Co., 1878), 1:10, 13.
    4.   Hodge, Systematic Theology, p. 60.
    5.   A letter written to a scientific periodical said
condescendingly about Einstein:

> Let us leave to the pure mathematicians the
> delightful occupation of rambling through
> wonderland with their imagination.   It would be
> unreasonable to deprive them of their mental
> relaxations and amusements in the land of dreams in
> which they have such ample scope for mental
> dexterity. All I maintain is that these dreams are
> entirely out of place in that branch of inductive
> thought called science.   Henry H. Howorth,

"Transcendental Premises in Science," Nature 107 (March 3, 1921):9.

6. New York Times, April 8, 1929, p. 4.
7. "What Life Means to Einstein. An Interview by George Sylvester Viereck," Saturday Evening Post, October 26, 1929, p. 17.
8. L. Franklin Gruber, "The Einstein Theory," Bibliotheca Sacra 79 (1922): 88.
9. New York Times, February 4, 1929, p. 26.
10. Ibid.
11. New York Times, April 25, 1929, p. 60.
12. Ibid.
13. "Einstein and Religion," Commonweal, October 29, 1930, p. 653.
14. Albert Einstein, James Murphy, and J. W. N. Sullivan, "Science and God," Forum 83 (June 1930), p. 373.
15. Albert Einstein, "Science and Religion," in Ideas and Opinions, trans. and ed. Sonja Bargmann (New York: Crown Publishers, Bonanza Books, 1954), p. 46. Originally published by the Conference on Science, Philosophy and Religion in Their Relation to the Democratic Way of Life, New York, 1941.
16. Einstein, "On the Generalized Theory of Gravitation," Ideas and Opinions, p. 342. Originally published in Scientific American, April, 1950, p. 182.
17. Einstein, "Principles of Research," in Ideas and Opinions, p. 226. Previously published in Mein Weltbild (Amsterdam: Querido Verlag, 1934).
18. Einstein, "The Problem of Space, Ether, and the Field in Physics," in Ideas and Opinions, p. 282. Previously published in Mein Weltbild.
19. Einstein, "On the Method of Theoretical Physics," in Ideas and Opinions, p. 274. This is the Herbert Spencer Memorial Lecture, delivered at Oxford, June 10, 1933.
20. "Principles of Research," p. 227.
21. New York Times, December 9, 1931, p. 6.
22. "God and Mathematics," Nation , September 27,1933, p. 342.
23. I Corinthians 13:12 (Today's English Verson).
24. Richard Tute, "Space-Time: A Link Between Religion and Science," Hibbert Journal 38 (1940): 269 - 70.
25. John Warwick Montgomery, Where Is History Going? (Minneapolis: Bethany Fellowship, 1972), p. 71.
26. For example, in the November 9, 1930 New York Times article entitled "Religion and Science," Einstein says: "The man who is thoroughly convinced of the universal operation of the law of causation cannot for a moment entertain the idea of a being who interferes in the course of events — provided, of course, that he takes the hypothesis of causality really seriously." Quoted from Ideas and Opinions, p. 39.
27. Bertrand Russell, "The Twilight of Science," Century Magazine, July 1929, p. 311.
28. Einstein, "Physics and Reality," in Ideas and Opinions, p. 322. Originally published in Journal of the Franklin Institute 221 (1936).
29. John Keats, Lamia, Pt. II, 11. 235 - 37.
30. "Science Needs the Poet," New York Times, December 21, 1930, Sec. 3, p. 1.

31.   See <u>New York Times</u>, April 4, 1921, p. 5.
32.   "Einstein Gets Us Guessing," <u>Harper's</u>, April 1929,
p. 653.

6.

Einstein on Kant, Religion, Science, and Methodological Unity

ROY D. MORRISON II

> Strange is our situation here upon the earth. Each of us comes for a short visit, not knowing why, yet sometimes seeming to divine a purpose.[1]

These are the opening words of Einstein's "Credo." They reflect a philosophical and religious perspective that is different from that of classical Christianity and from those Western philosophies of history which it has significantly influenced. For St. Paul and for St. Augustine, our situation on earth was unfortunate, but it was not "strange." We knew exactly why we were here. Our aim was to achieve eternal life, in another world, or through miraculous transformation, by worshiping and by obeying the supernatural, theistic god of the Judeo-Christian tradition. In the Genesis account of creation, in Aquinas, and in Hegel, history and the human situation are dramatized. There is a beginning, a middle development, and a cosmic or ontological culmination. Behind the scenes, and more or less inscrutable to humans, there is a divine person, or a divine principle that rationally and purposefully determines the course of history. Individual humans discover their ultimate purpose by faithfully and properly subscribing to the allegedly revealed purpose. The intent of the various dramatizing enterprises is to satisfy a typology, a cluster of potential needs that have been carefully, perhaps neurotically, cultivated in the consciousness of the Western world.[2]

The tranquil agnosticism reflected in our opening quotation from Einstein's credo indicates that he did not have a neurotic dependency on a miracle-working cosmic authority figure — and, he was not incapacitated by the possibility that there might be no ultimate purpose through which the human situation could be interpreted and made more palatable. In other words, Einstein was capable of acceptance. He was able to sustain a philosophical acceptance of the limited horizons of meaning and of moral expectations which characterize the human situation.

The intent of this essay is to celebrate the life and mind of Albert Einstein by arguing that there is a methodological unity underlying his approaches to religion and to science — and that serious, sustained consideration of his religious/theological reflection is just as obligatory and valuable for modern humanity as is consideration of his scientific enterprises. The method that has such revolutionary power to generate verifiable knowledge also has inescapable implications for philosophy, theology, and religion. The discussion will present what we believe to be Einstein's corrective to Kant's teachings with regard to the basic categories of thought and of reality. Also, Einstein's philosophy is briefly compared with that of Werner Heisenberg with regard to methodology and causal determinism.

Reflecting upon the history of methodological and conceptual strife between religion and science, Einstein makes the following observations:

> During the youthful period of mankind's spiritual evolution human fantasy created gods in man's own image, who, by the operations of their will were supposed to influence the phenomenal world. Man sought to alter the disposition of these gods in his own favor by means of magic and prayer.

> The main source of the present-day conflicts between the spheres of religion and of science lies in [the] concept of a personal God.[3]

Einstein acknowledges that the "idea of the existence of an omnipotent, just and omnibeneficent personal God is able to accord man solace, help and guidance."[4] Nevertheless, he advances two major kinds of reasons for rejecting the existence of a personal god. These reasons involve (1) causality and epistemological limits, and (2) theodicy and human values. Methodologically speaking, his first reason depends upon the category of causality and thus is situated in his metaphysics and epistemology. Though well aware that such a postulate could not be "proven," he believed that causality was a "rule," a "law of nature" with "absolutely general validity." He was so "imbued" with the ordered regularity that science discovers in nature that he had "no room left by the side of this ordered regularity for causes of a different nature."[5]

Einstein grounds his notion, or metaphysical postulate, of universal causal order in a religious attitude which he says is faith or something akin to faith. In his more epistemologically technical approach to the problem of a methodological warrant for this postulate, he rests his case on the assertion that certain working postulates are necessary (1) for thinking and (2) in order to avoid solipsism.

Einstein's second reason for rejecting the notion of a personal god presupposes a kind of human autonomy — the idea that humans have the right to make a rational critique of the inherited ideas of god, to question the moral behavior of an alleged god, and to reject theologies that are dehumanizing. Consequently, his second reason for rejecting the personal god is based on his response to the problem of theodicy — and it prepares us for his understanding of religion. Theodicy is the study of an irreducible question in theology and philosophy: if God is a centered consciousness, omnipotent, and absolutely righteous, why do suffering, injustice, and absurdity persist in the human situation? In his credo, from which we quoted earlier, Einstein gives his response with the following words:

> I cannot imagine a God who rewards and punishes the
> objects of his creation, whose purposes are modeled
> after our own — a God, in short, who is but a
> reflection of human frailty....It is enough for me
> to contemplate the mystery of conscious life per-
> petuating itself through all eternity, to reflect
> upon the marvelous structure of the universe which
> we can dimly perceive, and to try humbly to compre-
> hend even an infinitesmal part of the intelligence
> manifested in nature.[6]

This quotation expresses the core of his notion of
"cosmic religion" and also reflects his frequently cited
indebtedness to Spinoza. Elsewhere, Einstein is more
explicit in his explication of religion. He sees the reli-
giously enlightened person as one who has "liberated" himself
from the fetters of selfish desires and is preoccupied with
thoughts and aspirations that possess superpersonal value.
What is important is the overpowering meaningfulness of the
superpersonal content — "regardless of whether any attempt is
made to unite this content with a divine being, for otherwise
it would not be possible to count Buddha and Spinoza as
religious personalities."[7]
   For Einstein, science tells us what is; religion tells
us what should be. Religion does not deal with facts or with
relationships between facts. Rather, it deals only with
evaluations of human thought and action. Religion evokes
aspiration toward truth and understanding. Religion
generates "faith" in the intelligibility of the empirical
world. In turn, religion is nurtured by the reverence and
awe which accompany our discovery of the order and harmony in
the universe.[8] For Einstein, then, there is a single
attitude which lies at the base of religion, philosophy, and
science. This attitude, which is religious in the highest
sense of the word, motivates the striving for the highest
ethical ideals and the striving for the deepest possible
grasp of the intelligibility of the cosmos.
   In a manner that is necessarily brief and selective, we
now turn to the core of Einstein's methodology for philosophy, science, and religion. In his "Reply to Criticisms"
in the seventh volume of Paul A. Schilpp's Library of Living
Philosophers, he presents the following explanations and
arguments:
   1. One necessary prerequisite of scientific and pre-
scientific thinking is the distinction between "sense
impressions" (and the recollection of such) on the one hand
and mere ideas on the other.[9]
   2. He concedes that there is no evidence and "no such
thing as a conceptual definition of this distinction."[10]
Undeterred by the reproach that he is guilty of metaphysical
"original sin," Einstein designates this distinction as a
"category which we use" (emphasis added) in order that we can
function in the world of immediate sensations. In this
achievement lies the justification of the distinction.
   3. Einstein continues to state that:

We represent the sense-impressions as conditioned

by an "objective" and by a "subjective" factor.
For this conceptual distinction there also is no
logical-philosophical justification. But if we
reject it, we cannot escape solipsism. It is also
the presupposition of every kind of physical thinking.[11]

4. Insofar as physical thinking justifies itself, by its
ability to grasp experiences intellectually, we regard it as
"knowledge of the real."[12]

5. The theoretical attitude here advocated is distinct
from that of Kant only by the fact that we do not conceive of
the "categories" as unalterable (conditioned by the nature of
the understanding) but as (in the logical sense) free conventions. They appear to be a priori only insofar as thinking
without the positing of categories and of concepts in general
would be as impossible as breathing in a vacuum.[13]

The three preceding quotations indicate that Einstein
conducts a critique of Kant's epistemology and then makes his
own response to the epistemic problems revealed by David
Hume. From Hume, Kant had learned that certain concepts such
as causal connection dominate our thinking — though they
cannot be logically deduced from empirical data. The methodological question which Kant confronted, and attempted to
resolve, can be stated as follows: what is the epistemic
warrant or justification for the use of such concepts?
Einstein suggests that Kant could have made the following
two-step response: (1) thinking is necessary in order to
understand that which is empirically given; and (2) concepts
and "categories" are necessary as indispensable elements of,
or conditions for, thinking.[14]

Einstein then makes the extremely interesting observation that if Kant "had remained satisfied (emphasis added)
with this type of an answer, he would have avoided scepticism
and you would not have been able to find fault with him."[15]
Kant believed that he had proven the existence of synthetic
judgments a priori, judgments which are produced by reason
alone — and consequently have absolute validity. In one
passage, Kant states that "metaphysics consists, at least in
intention (his italics), entirely of a priori synthetic propositions."[16] On the one hand, Einstein denies the existence
of such judgments — as Kant formulated them.[17] On the other
hand, Einstein retains the methodological postulation of such
universal categories as causality. It is this hypothetical
postulation of the categories that are necessary for thinking
that Einstein designates as "the really significant philosophical achievement of Kant."[18] It is worth noting at this
point that Einstein's procedure and presentation are charac
terized by almost incredible intellectual economy when compared to the extended speculative intricacy of the Kantian
transcendental deductions.

One additional element of Einstein's methodology must be
cited here. He insisted that when two verified principles or
conclusions are contradictory, one must go back to the presupposition that causes the apparent contradiction and re-
place that presupposition with its negate. Einstein's methodology relentlessly follows this approach in developing
the special theory of relativity — as well as in treating the
problem of theodicy.[19]

Let us now summarize our discussion of Einstein's methodology for warranting the postulates that undergird philosophical and scientific thinking. (1) Einstein retains from Kant the indispensability of the distinction between objective and subjective reality. (2) He shares the notion that all knowledge depends upon and is limited by experience. (3) Like Kant, Einstein employs what F.S.C. Northrop calls the "two-termed epistemic correlation."[20] This procedure contains features of rationalism and of extreme empirical orientation.[21] The speculative or theoretically postulated factors are continuously correlated with empirically given data to produce knowledge about reality.[22] (4) Einstein does not posit rigidity or proof for the basic postulates. He regards them as free inventions of the scientific imagination, justified by indispensability, by their operative success in providing intelligibility, and supported, perhaps crucially, by an attitude that is "akin" to religious faith.

Continuing our summary, we find (5) that Einstein's capacity for "acceptance," as cited at the beginning of this paper, enables him to be satisfied with less ambitious moral, theological, and epistemic expectations than Kant. Hence Einstein is, in one sense, more agnostic than Kant. Moreover, we believe, he exhibits a more profound recognition of the limits of human knowledge than Kant. It is this writer's opinion that Kant falls into cognitive hubris in his attempt to provide a transcendental deduction for the rigid categories. Also, though Kant deserves his reputation for having designated the limits of human knowledge, he notoriously violates those limits himself with his notions of noumenal reality, freedom as causality, and with the so-called "practical" extension of reason. (6) Kant draws the limits of human knowledge at the boundary where his categories regulate the phenomenal world. Beyond this "transcendental horizon," there is no knowledge.[23] Yet, Kant proceeds with a dualistic, fragmented conceptualization of a noumenal world in which freedom functions as a unique kind of causality.[24] Einstein did not believe that moral responsibility presupposes freedom — and did not believe that human beings can transcend their causal nexus. Rather, he agreed with Arthur Schopenhauer that "a man can surely do what he wills to do, but he cannot determine what he wills."[25] Therefore Einstein was not under the Kantian compulsion to conceive of another world, of a practical reason, or to postulate a second causal order when formulating an anthropology.

Consideration of the underlying philosophy of Einstein's theories of relativity and their cultural impact leads to an important observation. The three theories for which Einstein is most famous are all titled theories of relativity. They are thus afflicted by a most unfortunate misnomer — a kind of semantic tragedy that perpetuates the myth that Einstein regarded everything as relative. The situation is compounded because in some quarters the term, "relative," is identified with that which is private, pluralistic, and subjective. The truth is that in the theories of relativity, many aspects of motion and many processes are designated as "relative." However, they are all relative to something, namely the speed of light in vacuo, which is an absolute. According to Einstein, the propagation of light and its absolute velocity are not subjective, or idealistic, or mathematical realities.

Rather, they are external, objective, physical phenomena and
they are independent of the human observer. The notion of
the relativity of Galilean (inertial) frames of reference and
of the expansion and contraction of measuring rods is def-
initely revolutionary. However, the most radical and most
valuable contributions of Einstein's philosophy of physics
and knowledge lie in the opposite direction — away from the
emphasis on relativity.

Nor does the principle of relative simultaneity give us
a universe in which all things are relative. Instead, it
saves us from a situation of chaotic, relativistic pluralism
constituted by each observer's immediate experience.   It
postulates a universe ordered by invariant physical laws.
The exact meaning of simultaneity is determined by that
invariant physical system, and private, relative experiences
must be corrected through precise reference to the behavior
of that independent system.

> The idea of invariance is the nucleus of the theory
> of relativity....To achieve objectivity of basic
> description, the theory must confer relativity upon
> the domain of immediate observations.[26]

Hence the decisive consequence of the principle of relative
simultaneity is not relativity but cosmic and epistemological
invariance.

Henry Margenau has observed that, "What makes the theory
of relativity extraordinarily important for philosophy is its
incisive answer to the problem of objectivity."[27] One might
add, first, that the theory of relativity emerges from
Einstein's methodology for science in general. Secondly, the
term "objectivity," has more than one meaning. It may indicate a quality of a method when the procedure is extremely
rigorous and an attempt is made to eliminate all anthropomorphic traits and yearnings. Objectivity may also connote
that kind of reality that is "really there" in the following
senses: (1) it will not go away through wishing or through
our refusal to recognize it; (2) it is physical, external to
our "minds," and independent of human observation for its
basic character and for its continued existence; and (3) it
is a kind of reality that is ordered by universal laws; that
is, laws that are invariant from observer to observer.

One can safely assert that the work of both Kant and
Einstein exhibits objectivity in the sense of methodological
rigor.   However, they differ significantly on objectivity
where the term refers to the status of existence or reality.
Kant is deeply convinced of the indispensability of the
principle of "objective validity." Without the operation of
this criterion, all of our thinking degenerates into mere
fantasy or phantasmagoria.[28]   Generally, for Kant, the
phrase, "objective validity," refers to perceptual objects
that are (1) located "outside" in space; (2) not produced by
the mind but by external, empirical causality; and (3) ordered by "universal and necessary" rules, forms, and categories.
Unfortunately, in seeking a transcendental deduction (justification) for the universality and necessity of these ordering factors, he was driven to locate them subjectively,
inside the mind.

As is well known, Kant asserted the existence and the

character of his forms and categories in "synthetic a priori propositions" which were guaranteed by the human mind. Several difficulties arise here: (1) technically, the objects of perception are not external since they are in space which is a "property" of our minds; and (2) the universal rules of reality are not really universal since they are only operative when a human mind is thinking or having an experience, that is, processing noumena through the subjectively located categories and thereby, in some sense, creating phenomena. A very powerful case can be made for the conclusion that, technically, Kant does not have an objective system of nature with invariant laws because the forms and categories do not exist independent of the human mind. Kant's residual idealism triumphs here and he is left with an intense but strangely crippled commitment to objectivity — as methodological quality and as metaphysical status. Kant wants to support the Newtonian view of the physical universe but he resorts to transcendental idealism in his methodology. It should be noted that, in practice, Kant apparently treated the physical universe as if his idealistic, metaphysical rules of reality were "out there" and had causal efficacy in a physical sense.

In his general scientific methodology, in his theories of relativity, and in his notion of objectivity, Einstein avoids oscillation between physical realism and any form of metaphysical idealism. This is another, important, aspect of his methodological unity. For Einstein, there is an explicit and continuously operative postulation of a physical world that is not given to us immediately through sense impressions. That postulated world is the intended referent of the enterprise known as theoretical physics.[29] This postulation of external, independent reality includes (1) the raw stuff of reality (energy, not noumena), and (2) the "categories" that relate, connect, order, and differentiate this physical stuff. Instead of the strangely abridged universality and objectivity of Kant, Einstein's system is based on philosophical realism and a thoroughly physical theory of nature. His position is not a naive realism, however, because the axiomatic structure of physics is not abstracted from sensory experience but is constructed as a free invention of the scientific imagination.

In the methodological context of the theory of relativity, objectivity connotes the quality of rigorous thinking and also the notion of "thereness" that is independent of a perceiving subject. Objectivity is designed into the formal structure of the theory itself. Finally,

> Objectivity becomes equivalent to invariance of physical laws, not physical phenomena or observations. A falling object may describe a parabola to an observer on a moving train, a straight line to an observer on the ground. These differences in appearance do not matter so long as the law of nature in its general form, i.e., in the form of a differential equation, is the same for both observers.[30]

Thus, we believe that Einstein's idea of objective physical reality, with independent, invariant physical laws, provides

a much-needed corrective to Kant's venerated solutions for
the problems of knowledge and of reality. In any case,
Einstein's science and his relativity theories do not relativize or subjectivize our knowledge. Instead, Einstein's
contributions provide monumental reinforcement for the
principles of invariance and of objectivity.

It is worth noting here that the disagreement between
Einstein's notion of causal determinancy and Heisenberg's
principle of uncertainty or indeterminancy is not merely a
matter of an interpretation of the category of causality.[31]
Einstein postulates an objectively real structure of physical
reality ordered by universal laws that we know only by tenta-
tively formulated theories that are subject to indirect
empirical verification. Werner Heisenberg, on the other
hand, speaks of three "layers" of reality: "the possibility
of or tendency toward an occurrence constitutes a kind of
reality — an intermediate layer of reality situated halfway
between the bulky reality of matter and the spiritual reality
of the idea or picture."[32] For Heisenberg, "nature itself"
is quite different from physical nature as it is encountered
by a physicist through his sensory equipment or detected by
instruments.

> In the beginning, therefore, for modern science, was the
> form, the mathematical pattern, not the material thing.
> And since the mathematical pattern is, in the final
> analysis, an intellectual concept, one can say in the
> words of Faust, "Am Anfang war der Sinn" — "In the begin-
> ning was the meaning."[33]

Thus, for Heisenberg, the uncertainty encountered at the
level of elementary particles is an ontological or metaphysical uncertainty. The uncertainty is inherent in the
structure of reality. A particular speculative value judgment is very powerfully operative here: the transcendent
values of primordiality or ultimacy must be identified with
something other than physical or material reality. This
judgment is philosophical but it is also religious.

On the other hand, for Einstein, the uncertainty encountered at the level of elementary particles was merely an
epistemological uncertainty. The reality that science seeks
to know and to describe is a physical/material system and the
allegedly universal categories of space, time, and causality
do not suddenly become inoperative at the level of subatomic
energy particles. Einstein does not deduce logical or causal
consequences from nonphysical entities and processes.
Neither does he require or speculate about a nonphysical
structure of reality in order to satisfy his particular sense
of cosmic reverence. Instead, he experiences transcendent
awe through contemplation of the order and harmony that the
physicist discovers in physical reality.

This very brief comparison is significant because it
reveals that the interpretation of the method, content, and
the referent of theoretical physics is determined by the
deeply held philosophical and religious perspectives of the
two thinkers. Einstein and Heisenberg both postulate an
unperceived structure of reality that is the object of our
cognitive enterprises. For Heisenberg, this final or
ultimate object is constituted by a kind of reality that is

radically different from physical reality. Therefore, he has
two radically different methods, a scientific one for
empirical physics, and an idealistic, speculative method for
the structure of existence that he regards as "truly real."
Since Einstein's quest is for a single kind of reality, he
operates within the limitations of methodological unity.

Einstein's relativity theories and their associated
conceptuality appear quite radical in contrast to Newton's
notion of absolute space. Einstein's theology also seems
quite radical in contrast to the traditional idea of a
personal, theistic god in Judaism and in Christianity.
Nevertheless, other significant figures in our culture conceive of god in nonpersonal, nontheistic terms. What such
figures share with Einstein is the removal of the category of
centered consciousness in their theological and philosophical
method. Thus, they all technically remove the category that
is necessary for the intelligibility of the idea of a
theistic god. Therefore, despite its apparent radicality,
Einstein's approach to theology possesses striking compati-
bility with certain basic elements in a theological trend
that runs from the beginning of the German Enlightenment
through the liberal wing of the Black Theology movement of
the 1970s.

All over the world individuals and institutions are
celebrating the centennial of Einstein's birth and his contributions to the methodology of natural science. In this
chapter, we have been attempting to show that his scientific
method does not stand in isolation. Rather, it is one major
element among others in a methodological and attitudinal
unity, which indicates the limits of human knowledge while
achieving almost incredible success in providing verifiable
knowledge about the external world and our transactions with
it. The scientific/epistemological element cannot, respon-
sibly, be separated from the elements and implications that
directly involve religion and our relation to ultimate
reality — however, it might be conceived.

Einstein goes beyond Plato, Aristotle, Newton and Kant;
he goes beyond the notion of the theistic father-god. Moreover, he never associated himself with the dialectical speculations of the counter-enlightenment German romanticists
(Schelling, Heleg, Heidegger and Tillich). Instead, Einstein
offers his own path to re-enchantment and to a profound but
nondogmatic religiosity that is highly compatible with scientific method, with critical philosophy, with a humanistically
oriented ethic, and with the human need for a sense of
ultimacy.[34] The method and the content of his continuously
verifiable achievements must now be taken into account by a
responsible theological enterprise. We believe that critical
philosophy of religion now has the task of reassessing our
religious heritage in the light of Einstein's work and its
methodological unity.

Notes

1. Albert Einstein, "Credo," in Living Philosophies: A
Series of Intimate Credos, edited by Henry Goddard Leach.
(New York: Simon and Schuster, 1931), p. 3.

2. In the twelfth century, Joachim of Fiore used the
Christian symbols of the trinity to develop a speculative

doctrine of history and eschatology. See the critical discussion in Eric Voeglin, The New Science of Politics (Chicago: University of Chicago Press, 1952). Concerning the reason for believing the fallacious eschatological drama, see pp. 111ff. Concerning gnosticism and the nonrecognition of reality as a matter of principle, see pp. 167 - 173. See also Paul Tillich, The Future of Religions (New York: Harper and Row, 1966), pp. 66 - 77.

3. Albert Einstein, Out of My Later Years (Totowa, N.J.: Littlefield, Adams & Co., 1967), p. 30f.

4. Ibid.

5. Ibid.

6. Einstein, "Credo," p. 6f.

7. Einstein, Out of My Later Years, p. 29.

8. Max Planck, Where Is Science Going? (London: Allen and Unwin Ltd., 1933), p. 13.

9. Albert Einstein, "Remarks to the Essays Appearing in this Collective Volume," Albert Einstein: Philosopher-Scientist, ed. Paul Arthur Schilpp (Evanston, Ill.: Library of Living Philosophers, 1949), p. 673.

10. Ibid.

11. Ibid.

12. Ibid., p. 674.

13. Ibid.

14. Gerald Holton, "'What, precisely, is "thinking"?' Einstein's answer," in Einstein: A Centenary Volume, ed. A. P. French (Cambridge, Mass.: Harvard University Press, 1979), p. 154ff. See also, Phillip Frank, "Einstein's Philosophy of Science," Reviews of Modern Physics 21, no. 3 (July, 1949): 349 - 355.

15. Einstein, "Remarks to the Essays," p. 678f.

16. Immanuel Kant, The Critique of Pure Reason, trans. by N. K. Smith (New York: St. Martin's Press, 1965), p. B 18.

17. Kant's critical idealism, and his a priori forms, categories, and propositions are criticized and often rejected by modern physics, psychology, logic and relativity. See a brief discussion in Max Jammer, Concepts of Space (Cambridge, Mass.: Harvard University Press, 1957), p. 137; cf p. 129f.

18. Einstein, "Remarks to the Essays," p. 678. See also Norman Kemp Smith's discussion of Kant and the hypothetical method of the natural sciences, in his A Commentary to Kant's Critique of Pure Reason (New York: Humanities Press, 1962), pp. xxxviii and 239, 239 n.1.

19. See F. S. C. Northrop, Science and First Principles (New York: Macmillan, 1931), p. 68f.

20. F. S. C. Northrop, The Meeting of East and West (New York: Macmillan, 1946), pp. 453, 442 - 454, 468 - 481, and 493.

21. Einstein, "Remarks to the Essays," p. 679f. See also Karl Popper, "The Revolution in Our Idea of Knowledge." Unpublished lecture given as part of the Frank Nelson Doubleday Lecture Series, April 11, 1979 at the Museum of History and Technology of the Smithsonian Institution, Washington, D.C. 20560. Copies are available from the Smithsonian. Popper cites ten principles that constitute Einstein's mature theory of scientific knowledge.

22. Albert Einstein, "Considerations Concerning the Fundaments of Theoretical Physics," Science 91, no. 2369 (May

24, 1940):   487.   Reprinted in Albert Einstein, Ideas and Opinions, pp. 323 - 335.

23.   T. Langan, The Meaning of Heidegger (New York: Columbia University Press, 1961), pp. 72 - 80.

24.   Kant, The Critique of Pure Reason, pp. B 472 and B 586.

25.   Einstein, "Credo," p. 3.

26.   Henry Margenau, "Einstein's Conception of Reality," in Albert Einstein: Philosopher-Scientist, ed. Paul A. Schilpp (Evanston, Ill.:   Library of Living Philosophers, 1949), p. 245.

27.   Ibid., p. 252.

28.   Kant, The Critique of Pure Reason, pp. B 240 - 41 B 246 - 47.

29.   Albert Einstein, The World As I See It (New York: Covici and Friede, 1934), p. 60.   See p. 156 in the 1935 edition published in London by John Lane, The Bodley Head. See also Albert Einstein, "Autobiographical Notes," in Albert Einstein: Philosopher-Scientist, ed. P. A. Schilpp (Evanston, Ill.:   Library of Living Philosophers, 1949), p. 81.

30.   Henry Margenau, "Einstein's Conception of Reality," p. 253.

31.   Werner Heisenberg, The Physical Principles of the Quantum Theory (Chicago: University of Chicago Press, 1930), pp. 3, 58; cf. pp. 62, 64.

32.   Werner, Heisenberg, "From Plato to Max Planck: The Philosophical Problems of Atomic Physics," The Atlantic Monthly 204, no. 5 (November 1959), p.   111.

33.   Ibid., p. 113, cf. p. 112.

34.   Albert Einstein, Cosmic Religion: With Other Opinions and Aphorisms (New York: Covici and Friede, 1931), pp. 48 - 51, 98.

7.

Einstein and African Religion and Philosophy: The Hermetic Parallel

CHARLES A. FRYE

The following chapter will delineate specific elements of African religion and philosophy which are either implicit in Einstein's work or which Einstein's work can assist contemporary Western society in seeing as reasonable and plausible descriptions of reality.

I.   The All is Mind; the universe is mental.

Fritjof Capra's The Tao of Physics has established that there are similarities in the wisdom teachings of the ancients and the theories of modern physics, particularly those introduced by Albert Einstein.[1] Significantly, Capra ends his book with the observation that:

> In modern physics, the question of consciousness has arisen in connection with the observation of atomic phenomena. Quantum Theory has made it clear that these phenomena can only be understood as links in a chain of processes, the end of which lies in the consciousness of the human observer.[2]

> [This] implies, ultimately, that the structures and phenomena we observe in nature are nothing but creations of our measuring and categorizing mind.[3]

Geoffrey Hodson's The Kingdom of the Gods provides a testimonial from Einstein that affirms this premise:

> I believe in God . . . who reveals Himself in the orderly harmony of the universe. I believe that Intelligence is manifested throughout all Nature. The basis of scientific work is the conviction that the world is an ordered and comprehensible entity and not a thing of chance.[4]

As is often the case, however, the limits of Western scientific speculation often form the center of Eastern reality — and also the reality of the Western mystic. Small wonder that seers like Hodson would be so anxious to quote Albert Einstein. Einstein validates and vindicates the worldview of Western mystics by giving it scientific support. It has been demonstrated that the traditional African and the European mystic share essentially the same world-view.[5] Yet, the African has not benefited from such an iden-tification, because since the seventeenth century, European mystics have been regarded, at best, as members of the lunatic fringe, and, at worst, as subversives.[6] Consequently, the African worldview — and there are enough continent-wide beliefs to talk about such a view — could probably also benefit from Einstein's scientific validation.[7]

For the African initiate/priest mind was/is All. This universal mind was often described as an all-pervading, sentient life force[8] which was stratified into levels of being,[9] differentiated by their rates of vibration.[10]

The purpose of human existence for the African was evolution toward godhood, achieved through the systematic expansion of the mental faculties; to include, what Jung would call, the unconscious.[11] This goal of human perfection was pursued through eugenics: the deliberate manipulation of the race through breeding for physical strength, moral inte-grity and mental dexterity.[12] Quite simply stated, persons who exhibited certain undesirable physical, moral, or mental traits, (i.e., defects) were not allowed to marry.

Africans had several systems of categories of the mani-fested life force or mind.[13] However, one of the more func-tional was described (some critics say created) by Janheinz Jahn. Surveying the beliefs of the Bantu, Jahn surmised that they had four major categories of being:

1.   Muntu: including the Monad (the Great Muntu), the gods, the ancestors, humans (the Bantu), and all those beings endowed with will.
2.   Kintu: literally, "thing"; Bintu "things," including animals.
3.   Hantu: time and place.
4.   Kuntu: modalities, such as joy, fear, and beauty etc. [14]

The common element in each category is the one life force, Ntu.

Jahn's categories do at least three things. First, they affirm, though obliquely, that humans are not at the top of evolution, but stand somewhere near the middle.[15] However, humans through the perfection of their will are potentially capable of exercising all the powers of the higher beings of the Muntu class. Secondly, these categories graphically demonstrate that animals (and presumably children before their eighth day of corporal life), time and space, other mental constructs, subjective states, and qualities are of less force than, and hence under the dominion of, humans. Put another way, neither fear nor time need have any valid control over human beings. Indeed, many Africans feel they can manufacture as much time as they need.[16] It is not re-garded as an absolute. Moreover, the concept of Hantu re-cognizes the inseparable bond of space/time.

Finally, these categories exhibit a remarkable consis-tency with quantum theory by suggesting the absence of any absolute distinction between mass and energy. Everything is a force, or in Einstein's terms, an energy field, an energy field imbued with intelligence. In categories 1 (Muntu) and 2 (Kintu), the field becomes more dense as objects and people

appear.   In categories 3 (Hantu) and 4 (Kuntu) and in the
upper reaches of Muntu, the field is more rarified. Africans
say: matter is congealed spirit; spirit is rarified matter.
Neither is absolute. They are manifestations of the energy
field, the life force, the universal mind.

II.   As above, so below; as below, so above.

> Everything in creation exists within you, and
> everything in you exists in creation. You are in
> borderless touch with the closest things, and, what
> is more, distance is not sufficient to separate you
> from things far away. All things from the lowest
> to the loftiest, from the smallest to the greatest,
> exist within you as equal things. In one atom are
> found all the elements of the earth.   One drop of
> water contains all the secrets of the oceans.   In
> one motion of the mind are found all motions of all
> the laws of existence.[17]

This passage by Kahlil Gibran encapsulates not only the
ideas of Robert Fludd[18] and Lewis Thomas[19] but also expresses
the primary assumptions of men like Einstein who seek to
plumb the minutiae of this world, the subatomic realm, for
answers to the mysteries of the universe as a whole.
Capra discusses the difficulty shared by atomic physicists and Eastern mystics alike when they try to objectively
describe this ultimate reality by using words.[20]
The ancients, including Africans, acknowledging this
difficulty and its danger, developed intricate symbols to
express this ultimate reality.[21]   Three examples of such
symbols which even today can be found carved on mundane
African artifacts or stamped out as bright prints on cloth
are   ◊ ,)1(, and ⚕ .
Superficially, ◊ means "love or marriage." But as a
segment of a chain or portion of a spiral it also resembles
the spiral formed by the table of elements, or the DNA
molecule. The symbol also seems, therefore, to represent
Life. Individual marriages then become reaffirmations of the
one life force and the Divine Love which sustains us all.
Next,)1(means "hatred or divorce."   But it also
suggests death or, at least, the absence of corporeal
existence, or the need for same.
Finally, ⚕ aside from looking somewhat like multiple
dollar signs, means "the curved line of falsehood written
over the straight line of truth."[22]   This symbol resembles
the ancient Hermetic caduceus which is still used as the
emblem of the medical profession. The curved lines of the
caduceus are two serpents entwined about a straight staff
which has a round head with wings extending from either
side. The serpents, one black one white, represent the human
descent into matter and the ascent back to spirit. The staff
represents the straight path of initiation into the
mysteries; the winged staff head represents clairvoyance,
among other things.   The curved line, the serpents, the
spiral, the chain links, when viewed thusly, become maya, the
illusion and falsehood of corporeal existence. The "straight
line of truth," which cuts through the ambiguity of earthly
existence, is thereby also death.[23]   And a symbolic death is

usually a part of every initiation.

The ancients did also use words to express the essential nature of things (cf. Jahn's "Nommo.") However, the words they used all had numerical equivalences. The ancients saw the keys to the truth of our existence manifested in the vibrations of color[24] and tone[25] and articulated by number. This cabalistic approach to numbers is essentially the same as the African approach: numbers were regarded as forces, as living entities, that could enhance or diminish the personal force of individuals, or groups of individuals, and which revealed directly — not by analogy or intellectual deduction, but directly — the secrets of the universe.[26]

The reliance by modern physics on mathematical equations is perhaps an unconscious acknowledgement of the important role which number has always played in transmitting matters of vital importance. However, since the West no longer takes a Pythagorean approach to numbers, such equations, including $E=mc^2$, only provide intellectual revelations.[27] What the ancients sought and achieved through number was the total, visionary experience of "preintellectual reality."[28]

The message has ever been the same, whether uttered by Dogon Shaman, Ogotemmeli[29] or Yaqui Shaman, Don Juan Matus,[30] direct experience is the only source of understanding. And where does one look? The secret is everywhere revealed, in every being and nuance of being. The human is himself a key, standing equidistant, in proportion, from the super-giant red star and the electron, "he is the mean between the macrocosm and the microcosm."[31] S/he is the measure of all things; a duplicate of the one.[32] Or as the ancients say: "to know about God, know all about man."

III. Nothing rests; everything moves; everything vibrates.

The African asserts that rocks are born from the earth, that they grow and are sentient.[33] He further asserts that the miner and the smith serve as midwife and priest in facilitating mineral evolution.

> The extraction of iron or gold is equivalent to an obstetrical operation before the end of pregnancy. By intervening in this underground embryology, the miner helps the earth deliver more rapidly. The blacksmith further accelerates this maturation in his furnace. The fire action is assimilated to a sexual act and the blast furnace to a uterus. The fusion and mixture of metals are regarded as a marriage precluding a new birth, that of the object.[34]

The alchemical practices of the smith are rooted in the principle that everything is already in motion: that everything vibrates. By introducing an entity with a higher vibratory rate, fire, to entities of lower vibratory rates, minerals, the masters of fire or smiths; hasten the reascent of matter to spirit.[35] Their assumption is that this is the object of evolution; the refinement of gross materiality to its original infinite, at-one-ment with the source of all.[36]

Atomic physicists, of course, acknowledge the principle of vibration and apply it in their study of subatomic "particles" by quickening the vibratory rates of those "particles" through bombardment. In the final analysis, however, the African assertion of the livingness of rocks is no more peculiar than the physicist's assertion that one of nature's fundamental "building-blocks," the neutrino, is "motion only" or "disembodied spin."[37]

IV. Everything is dual; everything has poles.

For the ancients, the fourth, or central item in a series of seven usually embodied the spirit of that series. Even in isolation, four was regarded, by the Bantu, for example, as a "perfect" number. Polarity as the fourth Hermetic principle certainly encapsulates the core of the message which these principles convey. And the unitive cultures which have drawn sustenance from these principles can probably be best understood in terms of those cultures' reliance on the idea of polarity.

A contemporary psychologist, Albert Rothenberg, has argued that Albert Einstein and other creative geniuses in the West have been able to accomplish so much because they have mastered what Rothenberg calls the art of "janusian thinking," coined after the multiple-faced Roman god of doorways and beginnings, Janus.

> In janusian thinking, two or more opposites or antitheses are conceived <u>simultaneously</u>, either as existing side by side, or as equally operative, valid, or true. In an apparent defiance of logic or of physical possibility, the creative person consciously formulates the simultaneous operation of antithetical elements and develops those into integrated entities and creations. It is a leap that transcends ordinary logic. What emerges is no mere combination or blending of elements; the conception does not only contain different entities, it contains opposing and antagonistic elements, which are understood as coexistent."[38]

He further suggests that nature's secrets are often locked in such contradictions, and cites as examples the discovery of the double helix model for DNA and Einstein's "happiest thought," the idea that a person falling from a roof is in motion and at rest at the same time which led to the formulation of the general theory of relativity.[39]

What Rothenberg really describes, however, is the central postulate of the Hermetic tradition. That postulate stresses that all contradictions can ultimately be resolved into complements. And in the meantime, they can, at least, be equally true.

The paradoxes unveiled by atomic physics are certainly comparable to those of African cosmology. For example, subatomic "particles" are at once destructible and indestructible because they can be divided again and again without ever becoming smaller.[40] Similarly, the dead African grandfather can incarnate in the newest infant(s) lending the force of his name and presence, while remaining an <u>undiminished</u> force in the realm of the ancestors.[41]

The world is always ending, even as it continues.

64 Charles A. Frye

V. Everything flows out and in; everything has its tides.

Continents rise and fall, as do nations, as do the
tides, as do plants, animals, and people in an endless cycle
of birth, growth, decay, and death.
There are instances when the world simply is not.
Einstein might call these intervals of periodic surges,
quanta or energy packets.[42] Africans might call these
intervals, pulsations or breaths (inhalation) of the
universal life force.[43]
For the African and physicist, alike, life or existence
is not a constant, unbroken stream of energy, but tides/waves
of rhythmic pulsations.
The recorded sound of the main artery of the human
mother as heard by the baby in the womb has been used to
quiet, soothe, and ultimately put crying infants to sleep.[44]
Remarkably, the sound is very much like the beat of drums
deep in the forest.
While listening to this sound, one is easily convinced
that the affinity that primitives everywhere (and perhaps
contemporary disco fanatics) have for the beat of the drum is
an affirmation of their continuing umbilical connection to
the womb of the Earth. The beat to which they dance (a
refraction of the cosmic dance) is the pulse of life.[45]

VI. Every cause has its effect; every effect its cause.

There are no accidents, as far as the African is concerned. The world is a web of interconnections. Everything
that happens anywhere influences everything else.
There are no innocent victims. "God is not mocked."
Every pain suffered is the result of some offense. Since a
single lifetime often cannot account for one's share of
afflictions, reincarnation is assumed.[46]
However, as atomic physicists have discovered, things
can get even more complicated when we view causes and effects
as rowed dominoes that can be flipped in either direction.
Perceived causes can then become merely the effects of other
causes which are themselves effects and so on.

Particle interactions can be interpreted in terms
of cause and effect only when the space-time
diagrams are read in a definite direction, e.g.
from the bottom to the top. When they are taken as
four-dimensional patterns without any definite
direction of time attached to them, there is no
'before' and no 'after,' and thus no causation.[47]

What Capra suggests in this passage and what the
principle of causation implies is that there is but one
cause, the source of all; and that everything else is but
effect masquerading as cause/effect dominoes.

VII. Gender is in everything.

The one becomes two. Two creates the manifested
universe. Inherent in the Duad is time (duration/history)
and space (separation/alienation). As a force, the
Duad'snature is creativity — and conflict.[48] Anatomically,

the Duad is manifested in the two hemispheres of the human
brain: the creative/intuitive/ideational right hemisphere and
the analytical/rational/operational left, and in the female
and male sex organs.[49]
Cosmically,

> We find that male and female are really the
> positive and negative aspects in nature — that is
> to say, that male is the positive or electrical
> quality and female is the magnetic, receptive or
> negative quality. When the two fuse, creation
> occurs.[50]

Eliade writes:

> We are here dealing...with a general con-ception of
> cosmic reality seen as _Life_ and consequently
> endowed with sex; sexuality being a particular sign
> of all living reality. Starting from a certain
> cultural level, the entire world — the world of
> nature as well as the world of things and tools
> made by man — is presented as endowed with sex.[51]

The "cultural level" to which Eliade eludes is that of
archaic or primitive man. Erny verifies this view. For the
African:

> The whole world, the natural world as well as the
> world of man-made objects, is seen as sexed; it is
> the fruit of procreation and is thus found
> symbolically in close relationship with the human
> child. Fish are the offspring of the rivers, wild-
> animals the offspring of the bush, roots the off-
> spring of the earth, 'lightning stones' the off-
> spring of thunder, speech the offspring of the
> mind. Cosmic reality is perceived as endowed with
> life, and sexuality is one of the essential
> elements of the living.[52]

Like the tail of the uroboric serpent, gender turns back
toward its own head, the principle of mind.[53] Gender is the
"glue" that makes the basic energy relation possible, whether
that relationship is manifested as galaxies merging, the
gravitational fields of planets attracting, men and women
wooing, or "quantum blackholes building electrons."[54]
It is relationships from the most basic to the more
elaborate which are the source of consciousness.

> Sensory and brain research have proved conclusively
> that only relationships and patterns of relation-
> ships can be perceived, and these are the essence
> of experience.[55]

Einstein's work has suggested that there is no absolute
standard of time and no solid objects. There are only energy
fields: interconnecting processes or relationships. Relationships are the source of consciousness. This consciousness seems to permeate all of creation, with apparent correspondences between events on every plane of being. (For

example, the sentient force that is Mars as planet, is also the imagery of and physical "fact" of iron, the color red, the musical tone c,[56] the number 3, and the mythic figure, i.e. psychological state, characterized by passions and activity.)

These interconnecting processes are all in constant motion often in apparent opposition, moving back and forth, in and out, every motion affecting every other (Karmic) motion in the orgiastic dance that is life.

CONCLUSION

Obviously this discussion has been less about Albert Einstein or even African philosophy than about the Hermetic tradition. But that is perhaps the real significance of Einstein: the occasion of the centennial of his birth makes such a presentation not only possible, but appropriate. The full significance of the implications of Einstein's work have yet to sink in/filter down, even with the scientific community.

Albert Einstein in his work and in his life has affirmed each of the principles we have discussed. He has done what any African initiate/priest would have done, if allowed. But Einstein, the atheist/saint, has affirmed the Hermetic tradition in the only way that the current age respects. In his own words:

> The religious geniuses of all ages have been dis-
> tinguished by [the] kind of religious feeling,
> which knows no dogma and no God conceived in
> man's image; so that there can be no church whose
> central teachings are based on it. Hence it is
> precisely among the heretics of every age that we
> find men who were filled with this highest kind
> of religious feeling and were in many cases
> regarded by their contemporaries as atheists,
> sometimes also as saints...

How can cosmic religious feeling be communicated from one person to another, if it can give rise to no definite notion of a God and no theology? In my view, it is the most important function of art and science to awaken this feeling and keep it alive in those who are receptive to it.[57]

Amen.

Notes

1. Fritjof Capra, The Tao of Physics: An Exploration of the Parallels Between Modern Physics and Eastern Mysticism (Boulder, Colo.: Shambala, 1975).

2. Emphasis added.

3. Capra, The Tao of Physics, pp. 300 and 277. See also: Jose A. Arguelles, The Transformative Vision: Reflections on the Nature and History of Human Expression (London: Shambala, 1975), pp. 158 - 159.

4. Geoffrey Hodson, The Kingdom of the Gods (Adyar: Theosophical Publishing House, 1970), p. 17.

5. Fela Sowande, "Black Folklore," BlackLines, (1971): 5 - 21.

6.   Anne Kent Rush, <u>Moon, Moon</u> (New York: Random House, 1976), pp. 53 - 57.   See also: Jose A. Arguellas, <u>The Trans-formative Vision</u>, pp. 36 and 78.

7.   See, for example: John S. Mbiti, <u>African Religions and Philosophy</u> (New York: Praeger, 1969); Pierre Erny, <u>Childhood and Cosmos: The Social Psychology of the Black African Child</u> (New York: New Perspectives, 1973); Janheinz Jahn, <u>Muntu: The New African Culture</u> (New York: Grove Press, 1961); Placide Temples, <u>Bantu Philosophy</u> (Paris: Pre-sence Africaine, 1959).

8.   Temples, <u>Bantu Philosophy</u>.

9.   V. C. Mutwa, <u>Indaba, My Children</u> (Johannesburg: Blue Crane Books, 1965), pp. 433 - 503.   See also: Jahn, <u>Muntu</u>, chapters 4 and 5.

10.   This belief has often been ridiculed by anthro-pologists as "animism."   See: James Hillman, <u>Re-Visioning Psychology</u> (New York: Harper and Row, 1975), p. 13.

11.   In recounting his initiation into the priesthood, Mutwa provides a detailed example: in <u>Indaba, My Children</u>, pp. 495 - 980.

12.   Edward W. Blyden, <u>African Life and Customs</u> (London: C. M. Phillips, 1908).

13.   Among Bantu peoples, for the Dogon see: Marcel Griaule, <u>Conversations with Ogotemmeli</u> (New York: Oxford University Press, 1965); for Zulus see: Mutwa, <u>Indaba, My Children</u>.

14.   Jahn, <u>Muntu</u>, chapter 4.   Also see: Hillman, <u>Re-Visioning Psychology</u>, p. 176, cf. Kuntu.

15.   Mutwa and Ogotemmeli are more explicit in this assertion.

16.   Mbiti, <u>African Religions and Philosophy</u>, chapter 3.

17.   Anthony R. Ferris, ed., <u>Spiritual Sayings of Kahlil Gibran</u> (New York: Citadel Press, 1962).

18.   See: Manly Palmer Hall, <u>Man: The Grand Symbol of the Mysteries</u> (Los Angeles: Philosophical Research Society, 1947).

19.   Lewis Thomas, <u>The Lives of a Cell: Notes of a Bio-logy Watcher</u> (New York: Bantam, 1975).

20.   Fritjof Capra, <u>The Tao of Physics</u>, chapter 3.

21.   For the African, knowledge is, literally, power, sacred power.   Not easily acquired, it should not be too freely disseminated.   See: D. T. Niane, <u>Sundiata: An Epic of Old Mali</u> (London: Longmans, Green and Co., 1965) pp. 41 and 92.

22.   These symbols and their meanings were taken from an exhibit of African art and culture ("African Art in Motion", 1974) hosted by the National Gallery of Art, Washington, D.C.,

23.   The horizontal line in the Cross is primary.   It represents the Feminine Principle; the fecund Earth; the source of creativity and unrealized possibility; the well-spring of syncretic religions and intuitive wisdom.   The horizontal line represents all-inclusiveness.   It is the great leveler, symbolic of both ends of the life cycle; the womb and the tomb.   The vertical line of the Cross is secondary.   It issues from the horizontal line as the son issues from his mother.   The vertical line symbolizes the Masculine Principle.

It is the phallus, the mound, the temple. It is
one-pointed. It is the sword of Reason which cuts
through the ambiguity of life. It pierces the veil
of darkness. It soars as on eagles' wings. As the
horizontal line represents infinite Be-ness, the
vertical line represents infinite striving.

The most ancient portrayals of the Cross
appear thusly: T with the male issuing from the
female — but still attached. T which is the
ancient Egyptian Tau is etymologically linked to
the Chinese Tao, the symbol of completeness,
representing the interplay between and integration
of the polar energies of (Feminine) Yin and
(Masculine) Yang.

The Cross in the later form + symbolizes among
other things, the separation of the two principles
and their continuing unity at the center. [Charles
A. Frye, "Higher Education in the New Age: The Role
of Interdisciplinary Studies," American Theosophist
(March 1977), 61 - 64.]

24. See: Roland Hunt, The Eighth Key to Color:
Self-Analysis and Clarification Through Color (London:
L. N. Fowler, 1965). See also: Corinne Heline, Healing and
Regeneration Through Color (La Canada, Calif.: New Age
Press, 1972).

25. See: Corinne Heline, Color and Music in the New
Age (Oceanside, Calif.: New Age Press, 1964).

26. Myth is not an early level of human development,
but an imaginative description of reality in which
the known is related to the unknown through a
system of correspondences in which mind and matter,
self, society and cosmos are integrally expressed
in an esoteric language of poetry and number which
is itself a performance of the reality it seeks to
describe. Myth expresses the deep correspondence
between "the universal grammar" of the mind and the
universal grammar of events in space-time. A hunk
of matter does not create a cosmos. The structures
by which and through which man realizes the
intellectual resonance between himself and the
universe of which he is a part are his
mathematical, musical, and verbal creations.
[William Irwin Thompson, At the Edge of History
(New York: Harper and Row, 1971), pp. 190 - 91.]

27. Manly P. Hall, The Secret Teachings of All Ages,
pp. LXX - LXXII.

28. See: Robert M. Pirsig, Zen and the Art of Motor-
cycle Maintenance (New York: Bantam, 1974).

29. Marcel Griaule, Conversations with Ogotemmeli.

30. Carlos Castaneda, The Teachings of Don Juan: A
Yaqui Way of Knowledge (New York: Ballantine, 1968); A
Separate Reality: Further Conversations with Don Juan (New
York: Pocket Books, 1972); Journey to Ixlan; The Lessons of
Don Juan (New York: Simon and Schuster, 1972); Tales of Power
(New York: Simon and Schuster, 1974); and The Second Ring of
Power (New York: Simon and Schuster, 1977).

31. Lincoln Barnett, The Universe and Dr. Einstein (New
York: Bantam, 1968), p. 22.

32. The science of number stands above nature as a way

of comprehending Unity. Numbers are the principle
of beings and the root of all sciences, the first
effusion of Spirit upon Soul.

Within the coordinate system that man
represents, the units of spatial definition become
the members of the body. A basic system of six
evolved that was related to man as proportional
extensions of his own anatomy. The finger (digit),
palm, foot, and cubit developed as the fundamental
units of measure. The height of man was taken as
six feet, the distance from his elbow to the tip of
his fingers as one cubit or six palms, the width of
a palm as four fingers, and the finger, or digit,
as six grains of barley placed side by side. The
foot was taken as four palms or sixteen fingers.

Star systems, stars, plants, animals, and man
manifest their own relative dimensions and their
respective magnitudes. Man's capacity to fill,
perceive, traverse, and define space increases and
decreases in direct proportion to his life cycle.
The scale of these activities lie within the
limited bounds of inches, feet, and miles, which
are all units of measure derived from man himself.

Six, as the first mathematically complete
number, not only expresses the proportional height
of man but also represents the basic directions of
motion and the surfaces of the cube. Thus 6 is
considered the number of the body (jism) and the
most appropriate proportional system to define or
extend it in space. [N. Ardalan and L. Bakhtiar,
The Sense of Unity, Chicago: Univ. of Chicago
Press, 1973, p. 25.]

   33. Mircea Eliade, The Forge and the Crucible (New
York: Harper and Row, 1971), chapter 4.
   34. Pierre Erny, Childhood and Cosmos: The Social
Psychology of the Black African Child, p. 34.
   35. Eliade, The Forge and the Crucible, chapter 8.
   36. Max Heindel, The Rosicrucian Cosmo-Conception
(Oceanside, Calif: Rosicrucian Fellowship, 1909).
   37. L. C. Beckett, Movement and Emptiness (Wheaton,
Ill.: Theosophical Publishing House, 1968), pp. 52-59. See
also: Capra, The Tao of Physics, pp. 225 - 45.
   38. Albert Rothenberg, "Creative Contradictions,"
Psychology Today, June 1979, pp. 55 - 62.
   39. Ibid. See also: Jose A. Arguelles, The Transformative Vision, pp. 5 - 27.
   40. Capra, The Tao of Physics, p. 78.
   41. Erny, Childhood and Cosmos, pp. 107 - 16. See
also: Jahn, Muntu, pp. 110 - 11.
   42. Capra, The Tao of Physics, p. 67.
   43. See also: Alan Watts, The Book or the Taboo
Against Knowing Who You Are (New York: Coller, 1966), pp. 10
- 17.
   44. Dr. Hajime Murooka, "Lullaby From the Womb" (New
York: Capitol Records, 1974).
   45. Capra, The Tao of Physics, pp. 225 - 45.
   46. Erny, Childhood and Cosmos, pp. 99 - 146.
   47. Capra, The Tao of Physics, p. 186.
   48. See: Manly P. Hall, The Secret Teachings of All

Ages, pp. LXX - LXXII.

49. See: Julian Jaynes, The Origin of Consciousness in the Breakdown of the Bicameral Mind (Boston: Houghton Mifflin, 1976). Cf. Hillman's "soul" and "spirit" discussions in Re-Visioning Psychology, pp. 68 - 69.

50. Vera Stanley Alder, The Finding of the Third Eye (London: Rider, 1968), p. 74.

51. Mircea Eliade, The Forge and the Crucible, p. 36.

52. Pierre Erny, Childhood and Cosmos, p. 37.

53. Erich Neumann, The Origins and History of Consciousness (Princeton: Princeton/Bollingen, 1954).

54. Bob Toben, Space-Time and Beyond, New York: Dutton, 1975, p. 98.

55. Paul Watzlawick, Janet Helmick Beavin, and Don D. Jackson, Pragmatics of Human Communication: A Study of Interactional Patterns, Pathologies, and Paradoxes (New York: W. W. Norton, 1967), p. 27.

56. Alder, Finding the Third Eye, p. 99.

57. Albert Einstein, Ideas and Opinions (New York: Bonanza Books, 1954), p. 38. See also: Mutwa, Indaba, My Children, pp. 495 - 98.

Part IV

Metaphysics

8.

The Nature of Causality and Reality: A Reconciliation of the Ideas of Einstein and Bohr in the Light of Eastern Thought

RICHARD DOBRIN

Einstein believed in an external physical world independent of the perceiving subject.[1] Sense perception (that is, the result of experiment) gives us indirect information about the external world. We are only able to grasp this physical world by speculative means. That is, our physical theories are products of the mind, and not a summary of sensory experience. Our theories, our notions of the external world, must ultimately agree with sense experience, and be modifiable as the perceived facts require. The axiomatic structure of physics will never be final, as physical laws are only our notions of an external physical reality, and not that reality itself. Our notions of the external world are not that reality and can never be that reality. The physical world is a definite world, a world of complete causal relations. But Niels Bohr's and Werner Heisenberg's notion of indeterminacy is totally at variance with a complete causal interpretation of nature that demands that all physical quantities be simultaneously measurable in order to assure strict predictability of physical phenomena.[2] This chapter suggests that these differing views can be reconciled in the light of Hindu Tantric cosmology.

At the fifth Solvay Conference in Brussels in October 1927 Einstein and Bohr agreed to explore the issue in depth. This problem, the question of whether quantum theory provides the fullest possible account of microphysical phenomenon, or as Einstein maintained, a more detailed account of these happenings could be developed, was at that time, and has remained, a fundamental problem in the foundations of physics.[3] Starting with a reexamination of the thought experiments used by Heisenberg to formulate the indeterminacy relations, Bohr and Einstein invented more and more elaborate experiments to bring their points forward. Einstein aimed at disproving the indeterminacy relations through the use of these thought experiments, for if this could be done, he would refute Bohr's assertion of the incompatibility of simultaneous energy-momentum and space-time descriptions of physical phenomena, and with it his entire theoretical argument.

Einstein proposed the following experiment: consider a box containing a certain amount of radiation, and a clock mechanism which can open a shutter at a preset time for a very short interval, so that only one photon is released.

The mass of the photon could be obtained by weighing the box before and after the release of the photon. The energy of the photon is calculated from the mass by the equation: $E = mc^2$. Here, E is the photon energy, M is the photon mass, and c is the velocity of light in a vacuum (3×10^8 meters per second). Since the energy is exactly determinable and the time the photon is emitted can be determined as accurately as possible by decreasing the time the shutter is opened, the relation $\Delta E \; \Delta t \geq h$ is refuted since $\Delta E = 0$ and Δt can be made arbitrarily small.

After spending a sleepless night, Bohr replied with a thought experiment of his own which used Einstein's own general theory of relativity to disprove his arguments of the previous day. Bohr suspended the box containing the photons from a spring balance. A pointer indicates the position of the box and therefore its mass. When a photon is released, the box rises. However, the general theory of relativity states that the rate a clock runs changes as the clock moves toward or away from a gravitational mass, in this case the earth.[4] After the departure of the photon, the new weight and therefore mass of the box must be determined. The energy of the photon leaving the box is obtained from the equation: $\Delta E = \Delta mc^2$; where Δm is the difference between the mass of the box before and after the photon left, c is the velocity of light, and E is the energy of the photon. The precision to which we determine Δm, depends upon the accuracy of the balancing procedure. The more precise the measurement, the longer the balancing interval, the greater the displacement of the box in the gravitational field, and hence the greater the error in reading the clock arising from the rate change. It can be shown from this experiment that $\Delta E \; \Delta t > h$.

At this juncture Einstein conceded the logical consistency of the Heisenberg relations and of Bohr's point of view. However, his acquiescence was only an admission of the epistemological correctness of these relations. That is, these relations combine both experiment and the use of reason in their formulations. They must not be considered ontological in character, that is, they do not refer to the subject matter of scientific knowledge, the external world of nature which does not depend on its relationship to the perceiver. Since he maintained that this world is of a definite character, Einstein refused to give up his adherence to strict causality and still maintained that a completely causal description of microphysical phenomena could be given, one in which the energy, momentum, position, and time of measurement could be simultaneously obtained for a given physical system. Einstein now changed his method of approach in these debates and attempted to refute quantum theory on the grounds of inconsistency in a theoretical paper written with B. Podolsky and N. Rosen. The authors set forth the following conditions concerning physical reality:

1. For a physical theory to be complete "every element of physical reality must have a counterpart in physical theory."

2. "If without in any way disturbing a physical system, we can predict with certainty the value of a physical quantity, then there exists an element of reality corresponding to this physical quantity."[5]

A thought experiment described in this paper will be

presented have in a simplified manner by considering a system
consisting of two particles denoted 1 and 2. It can be shown
that after collision the measurement of the momentum of
particle 1 allows us to predict with certainty the momentum
of particle 2. Furthermore, a measurement of the position of
particle 1 allows us to predict with certainty the position
of particle 2. Hence the position and momentum of particle 2
have a physical reality from reality criterion (condition
2). However, quantum mechanics says we cannot measure both
the position and momentum of the same particle simultaneously; hence, quantum theory cannot encompass both elements
of the same physical reality and it is therefore incomplete.

Bohr responded by saying that the choice of measuring
either the position or momentum of particle 1 involves a
choice of mutually exclusive experimental procedures. If one
chooses to measure the position of particle 1, the interaction of particle 1 with the measuring apparatus precludes
any subsequent accurate measurement of the momentum of particle 1 and, therefore, the momentum of particle 2. Similarly,
measuring the momentum of particle 1 precludes the measurement of the position of particle 2, even though they do not
interact. Hence, the Einstein, Podolsky, Rosen (EPR) argument is disproven. Bohr's underlying thesis in disproving
the EPR argument was that the objects under investigation and
the measuring apparatus cannot be separated. That is, the
description of a state of a system is a relationship between
the particles under investigation and the measuring device.

The thesis of this chapter is that these opposing view-
points can be reconciled in light of Eastern thought. In
particular, Tantra has been chosen to express the common
aspects of many Eastern systems of thought because: (1) it
incorporates several earlier metaphysical systems within its
approach to achieving transcendence, and (2) it is deeply
involved with astronomy, cosmology, and the nature of the
physical and subphysical world.

In the Hindu Tantric cosmology, Brahma, the formless
one, who exists beyond the created materium, in the initial
act of creation manifests itself as a dualistic creative
force.[6] The portion of Brahma manifest in creation is called
Sakti (female energy), who is also Prakrite (the cosmic force
of nature) the impelling energy and prime mover of creation.
The portion of Brahma whose nature is static and exists in
the transcendental plane beyond creation is known as Siva
(male energy), which is also Purusha (cosmic consciousness).
This fundamental dualism pervades all of creation as male and
female, static and active, consciousness and nature. However, this dualism is illusory as each pair are two aspects
of one fundamental inseparable principle.

To the Tantric yogi, sound or vibration is the seed from
which all of creation springs. First to issue from the bosom
of Brahma is unmanifest pravana sound, which is an aggregate
of all existing sounds, and as such gives birth to the cosmic
process itself. Pravana sound gives rise to anahata-dhvani
or unstruck sound (that is, sound without vibration). The
immensely powerful anahata-dhvani can create, destroy,or
remold the structure of the creation. This single primary
sound-before-sound subdivides itself into fifty mantrika
sounds which evolve and permute into the manifold sound forms
which are the matrices for all created objects. That is, the

Tantrics believe that underlying each created manifestation, be it thought or material objects, is a sound concentration or matrix. Sound creates light. Light is seen as sound at a particular frequency.

There are many levels of creation manifested in this process of subdivision, many distinct universes of which this material universe we live in is but one. All things are created of the substance of Brahma as Sakti, and hence the individuality of things and their apparent substantialness are but cosmic illusion of maya. This shifting perspective of names (nama) and forms (rupa) in which we live is but the cosmic dance of the Divine Being (lila), and all duality is no more than the illusion of maya.

To achieve transcendence is the aim of all Eastern metaphysical systems. Freedom from the cycles of reincarnation and achieving unity with the unmanifested creative force are sought in Tantra through meditative processes and the observation of ritual. A key element in this procedure is the creation of Tantric visual representations of cosmic forces or energy patterns called Yantra. The Tantric artist who creates these representations engages in an elaborate ritual of meditation and visualization of a particular aspect of the Deity before rendering the corresponding visual image. The pattern represented may be in the form of simple geometrical figures such as straight lines, triangles, or circles, or somewhat more complex structures. The Yantra is a representation of a particular primordial force and is often referred to as an energy pattern or power diagram. This revealed image of an aspect of cosmic structure is a visual representation of specific vibrational energy patterns. In Yantras we have the underlying reality of cosmic structure portrayed in terms of the concepts of infinity, time, space and the interaction of polarities. The Tantric yogi, in meditating upon a specific Yantra, establishes an inner intuitive connection with the underlying vibrational pattern, and therefore expands his identification with the creative deity. Cosmological formulations, based on both observation and intuitive insight, were visually represented.

Science played an important role in the Tantric system. Aspects of science that had a practical,ritualistic interest to the Tantrikas were freely adopted from the extensive body of ancient Indian scientific knowledge. In this way astronomy, astrology, and notions of molecular and atomic structure became incorporated into the Tantric framework. Our object here is to analyze the views of Einstein and Bohr in the light of this system.

For Bohr, the quantum mechanical state had become a relational conception. That is, there existed an unanalyzable wholeness in the interaction of the measuring device and the microobject under examination. Aaga Petersen quoted Niels Bohr in saying, "There is no quantum world. There is only an abstract physical description. It is wrong to think the task of physics is to find out how nature is. Physics concerns what we can say about nature."[7] Hence, there is no physical reality, no quantum mechanical objects. The world quantum characterizes a type of description, not a description of quanta. There are only the results of our measurements, and our theories based on them. They do not describe "reality," and it is meaningless to talk of such a concept. It is meaningless and unknowable. There are no separate objects of measure that we can talk about independent of our measuring instruments. Hence, there is no way one can discuss an "independent" physical reality, an ontological conception divorced from epistemological considerations. For Einstein, this ontological reality existed, whose existence should reflect a strict causality in our physical theories.

These two extreme points of view on the nature of reality appear irreconcilable in the Western scientific framework of which they are a product. Is it reasonable to attempt their unification in terms of Eastern thought, a system in which polar opposites are considered two inseparable portions of a unified whole?[8] The rationale for this venture is clearly demonstrated if one examines the scientific achievements of ancient India, and compares them to our current physical knowledge of the microscopic and macroscopic worlds. The modern physicist interacts with nature by means of instruments, and the data obtained is used to expand man's perspective of the physical world. An intellectual-intuitive construct which unifies and offers a simple explanation for disparate phenomena is called a physical theory. The twentieth century has seen the introduction of theories that have vastly expanded our knowledge of the physical universe. Among these are the nuclear nature of atoms, the wave description of particle behavior, the big bang theory of cosmic evolution, and Einstein's general theory of relativity.[9]

In both East and West, measuring instruments have been used to obtain an accurate description of planetary motion. However, the Indian view of microscopic physical structure and cosmological events has been formed using man himself as the measuring instrument. Detailed descriptions of atomic and subatomic particles and the structural arrangement by which they form physical matter, and an elaborate cosmology based upon an expanding universe (as well as its eventual collapse, reintegration, and reexpansion, — the days and nights of Brahma) have been made on the basis of personal inner exploration. This description of cosmic birth not only speaks of our universe but an infinity of others, which evolve together with ours in the process of the subdivision of primal matter.

Tantric paintings from the eighteenth century showing the evolution of many different universes from Brahmananda (the cosmic egg) are reminiscent of Penrose diagrams. Figure 1 is a Penrose diagram showing the multiplicity of universes available to a traveler through an electrically charged black hole. The parallels between the intuitively obtained Tantric cosmological representation, and that derived through contemporary astrophysical methods is quite striking. Other paintings show primal atoms, that is, the fundamental constituents of subatomic particles in the physical universe. Contemporary physicists have also postulated the existence of subatomic particles called quarks which are said to be the constituents of hadrons, examples of which are neutrons and protons.

On the basis of the examples discussed above, and the many other parallels that can be shown to exist between contemporary physics and Indian thought, I consider it reasonable to examine the Western scientific debate between Bohr and Einstein in terms of the cosmological viewpoint of

Figure 1

A Penrose diagram of the Kerr solution for the rotating black hole. The diagram repeats infinitely into the past and future.

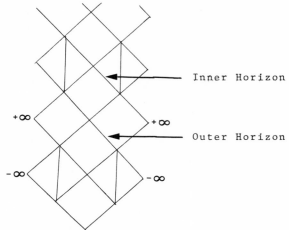

Inner Horizon

Outer Horizon

ancient Eastern science.

For Bohr, the observed and the observer cannot be separated. This is also true in Indian metaphysics where the existence of distinct objects and entities in this realm of names and forms is illusion of maya. Even if we assumed that the electron and other subatomic particles have a physical distinctness (as Max Born the well-known physicist would have us believe), as we probe deeper to find their constituents, we discover the nonmaterial atoms described in the Vaiseshika and Jaina schools of thought, the basic constituents underlying subatomic particles. Some of these may very well be our presentday quarks. Probing beneath this level of non-material atomicity, we find matrices of nonmaterial sound vibrations. The ability to carry out physical measurements, if it does not break down at the level of quark-atoms certainly collapses at the level of cosmic sound. At this causal level the Tantric yogi makes his "measurements" by meditation, by becoming one with the cosmic process under "study." The measuring instrument, the yogi, in his attempt to internally apprehend a phase of the creative Deity, and this "phase" itself, cannot be separated at the moment of apprehension or "measure." If the Tantric artist paints his conception of a cosmic force, this representation is only as accurate as the measuring instrument, that is the purity of the receptive channel of the Tantric artist. Hence in Tantra, as in physics, there is no way to ascribe a separate reality to the object under scrutiny, independent of the observer.

Does this mean that the ontological ideas of Einstein would find no parallels in Eastern thought? Far from it. For there is a reality, an absolute reality beyond time itself, but it is not a physical reality. The one original primal <u>pravana</u> sound is the aggregate for all existing sound and the potential for all of creation, the ultimate sum of archetypes. The fifty mantrika sounds into which the pravana sound subdivides have embedded within them the possibility of all new forms. These sounds are a timeless reality. A realization of this level of reality can only be approached through meditation and not physical measurement. Einstein also believed that our notions of reality do not come from measurement but through the "mental" which can be interpreted as the intellectual-intuitive process.

The nature of all creation is duality. The inclusion of the apparently contradictory ideas of Bohr and Einstein in the metaphysics of ancient India can be thought of as a kind of cosmic complementarity, that is, apparently irreconcilable descriptions, yet both necessary for a complete description of cosmic events.

Duality will always present itself in conflicting forms. Can we ever totally unite these oppositional elements into a unified whole? David Bohm, theoretical physicist and philosopher of science offers us a possible answer in his picture of creation.[10] The cosmos is like an onion, peel off one layer and another one lies beneath. The first layer is classical mechanics, peel it off and we find the causative level of the atom. Each level has a "reality" of its own independent of the levels beneath responsible for its behavior. For Bohm and Krishnamurti, reality, the stuff of creation, is relative and dualistic because there is always the observed and the observer.[11] There is no place within creation where this situation does not exist. The perceived requires a perceiver, even if the perception is intuitive. Truth, the nondualistic perception, can be obtained only if one is beyond creation, beyond reality. The liberated spirit beyond creation can become one with the totality of creation, be creation, and therefore perceive all in a nondualistic state. In this sense pravana and mantrika sounds are dualistic and in reality, for an aspirant to perceive them he must be separate from them.

To summarize, the positions of Bohr and Einstein form a complementary picture from the Eastern point of view. Bohr asserts that there are no distinct,isolatable, "real" elements of the measurable universe. Eastern thought maintains that creation is the dynamic interplay of the shifting elements of maya. Maya is the creation of Brahma, as Sakti, hence all elements are inseparable parts of a single totality. Einstein asserts that there exists a definite and causal reality, beyond our measurements and theories. That causal reality can be viewed as the mantrika, anahatadhvani or pravana sounds. Most fundamentally, the causal truth is Brahma, the formless consciousness, immeasurable and unfathomable, who from beyond duality creates duality, the basis of which is the Platonic archetypes which the ancient Indians see as fundamental vibratory sound forms.

Notes

1.   W. Heisenberg, Physics and Philosophy, (New York: Harper and Row, New York, 1958); H. Morgenau, "Einstein's Conception of Reality," in Albert Einstein: Philosopher-Scientist, ed. Paul Arthur Schilpp (Evanston, Ill.: Library of Living Philosophers, 1949).

    2.   M. Jammer, The Philosophy of Quantum Mechanics (New
York: John Wiley, 1974); F. S. C. Northrup, "Einstein's Conception of Science," in Albert Einstein: Philosopher-Scientist.
    3.   Jammer, the Philosophy of Quantum Mechanics; G.
Stent, "Does God Play Dice?" The Sciences, vol. 19, March
1979; N. Bohr, "Discussions with Einstein on Epistemological
Problems in Atomic Physics," in Albert Einstein: Philoso-
pher-Scientist.
    4.   N. Bohr, "Discussions with Einstein on Epistemo-
logical Problems in Atomic Physics."
    5.   A. Einstein, B. Podolsky, and N. Rosen, "Can the
Quantum Mechanical Description of Reality be Considered Com-
plete?" Physical Review 57 (1935): 777 - 80; see Morgenau,
"Einstein's Conception of Reality."
    6.   H.   Zimmer, Philosophies of India (Princeton:
Bollingen Series, 1951); A. Mookerjie, Tantra Art (New York:
Wittenborn, 1972); A. Mookerjie, Tantric Asana (New York:
Wittenborn, 1972); A. Mookerjie, The Tantric Way (Boston:
New York Graphic Society, 1977); A. Mookerjie, Yoga Art
(Boston: New York Graphic Society, 1975); P. Rawson, The Art
of Tantra (Boston: New York Graphic Society, 1973); P.
Rawson, Tantra: The Indian Cult of Ecstasy (New York:
Bounty Books, 1972).
    7.   Jammer, The Philosophy of Quantum Mechanics, p.
204.
    8.   W. Chan, The Way of Lao Tsu (Tao-te Ching)
(Indianapolis: Bobbs-Merrill, 1963).
    9.   W. Kaufmann, The Cosmic Frontiers of General Rela-
tivity (Boston: Little, Brown, 1977).
    10.   D. Bohm, Causality and Chance in Modern Physics
(Philadelphia: University of Pennsylvania Press, 1971).
    11.   I. Krishnamurti, Truth and Actuality (New York:
Harper and Row, 1978).

9.
Einstein and the Limits of Reason
RICHARD FLEMING

Philosophers have struggled through the ages attempting to clarify the roles played by experience and reason in human existence. Their attempt, essentially, has been one of sorting out the necessity and limits of each for human understanding. In what follows I will outline Einstein's view of the limits of reason in our rational investigations. I will do so by presenting the fundaments of his metaphysical and ethical doctrines, focusing attention on his view of the general nature of rationality which — surprisingly — proves to encompass the nonrational elements of human existence.

The early twentieth century found a large portion of the Anglo-American philosophical community antagonistic toward metaphysical investigations. Many of these antagonists to metaphysics were to rally around the philosophical doctrine of logical positivism which had at its heart the verifiability principle. This principle calls for the rejection of all statements and philosophical speculations that cannot be verified empirically or at least logically.

The logical positivist position was too empirically radical and too overly afraid of metaphysics to suit the tastes of Einstein.[1] This fear of metaphysics he held to be a great danger to the discipline of philosophy and viewed it as "the counterpart to that earlier philosophizing in the clouds, which thought it could neglect and dispense with what was given by the senses."[2]

Thus Einstein agreed with the philosophical empiricists who held that the notion of a "real external world" must rest on sense impressions. However, he did not believe that sense impressions alone could be taken as establishing the world. He argued that the first step in the setting of a 'real external world' is the formation of the concept of bodily object and of bodily objects of various kinds. This concept of bodily object, to be sure, is created from the multitude of sense experiences that repeatedly occur to us. But from this repetitious experience of sense impressions, we give a meaning or a use to the concept of bodily object. Einstein was careful, however, not to identify or equate this concept of bodily object with that of the totality of sense impressions it refers to, for the former is to be seen as an arbitrary creation of the human mind.

Einstein held that it is important that such metaphysical concepts not be given an a priori epistemological basis. These concepts have reference to sensible experience, but they are never, in a logical sense, deducible from them. "For this reason," he stressed, "I have never been able to understand the quest of the a priori in the Kantian sense. In any ontological question, the only possible procedure, is to seek out those characteristics in the complex of sense experiences to which the concepts refer."[3] Philosophers must, according to Einstein, stop creating harmful a priori categories and begin giving credit to the free creative ability of the mind.

Further, Einstein did not believe it justified and found it to be misleading and erroneous to veil the logical independence of metaphysical concepts from sense experience by advocating a theory of abstraction or induction. The more helpful and ultimately essential way to deal with metaphysical concepts and their connection to experience is to represent those concepts and the connected theorems as logically deduced from a narrow basis of fundamental, although freely chosen, axioms.[4] Thus rather than holding to an inductive procedure, we must admit the existence of an axiomatic basis and deduce our world from it. The axioms, of course, cannot be held to have any certainty or truth attached to them. While this in no way invalidates their use, it does present us with a constant lack of certainty that accompanies our theoretical investigations. The need for certainty, however, Einstein found to be a harmful desire.

> Sense perception only gives information of this external world or of 'physical reality' indirectly, we can only grasp the latter by speculative means. It follows from this that our notions of physical reality can never be final. We must always be ready to change these notions . . . in order to do justice to perceived facts in the most logically perfect way.[5]

Thus a continued refinement and at time a total change of our metaphysical ideas is important if we are to understand the external world.[6]

The fundaments of Einstein's metaphysical view involve, therefore, a limit to our rational comprehension of the external world. That is, certain fundamental concepts are freely chosen and — while being importantly connected with experience — they in no way are equivalent with or inductively derived from experience. We must, therefore, make certain fundamental choices concerning our concepts of the external world, and it is from these that we rationally derive and achieve an understanding of experience. The rational comprehension that we achieve, therefore, is clearly limited by the conceptual and axiomatic choices that we make, for such choices are speculative and free creations of the human mind.

This need to make free conceptual metaphysical choices in order to comprehend the external world plays a key role in Einstein's rejection of quantum mechanics. He said,

> What does not satisfy me in that theory, from the standpoint of principle, is its attitude towards

that which appears to me to be the programmatic aim of all physics: The complete description of any (individual) real situation (as it supposedly exists irrespective of any act of observation or substantiation).[7]

In fact, Einstein himself seemed to know that for all his examples and all his arguments, the conflict between him and his attempt to reach a unified field theory and the quantum physicists depended precisely upon "which direction one believes one must look for the future conceptual basis of physics."[8] This outlook in turn depended on one's metaphysical framework and understanding of the relationship between our concepts and experience. Quantum mechanics, for Einstein, lacked any type of consistent metaphysical doctrine, due mainly to its inherent reliance on positivistic principles, and thus its emphasis on the future conceptual basis of physics was naturally in conflict with Einstein's and would always be so.

I turn now to Einstein's ethical doctrines. Is it possible, David Hume asked, to arrive at a value statement (that is, a statement of what one ought to do) from concrete facts, from what is the case? Hume emphatically denied that such a derivation of an "ought" from an "is" is logically possible, and many philosophers since Hume have followed suit. Some, however, like the logical positivists, carried his point to even greater extremes and argued that ethical claims are not only not justified by what is but are in fact meaningless.

Einstein was one who found Hume's arguments to be persuasive. He agreed, for instance, that reason and scientific methods can provide us with no more than the relationships between facts and their conditions with respect to each other. Thus, he found it hardly necessary to argue that our ultimate goals, ethical or otherwise, and our desires to reach such ends must derive from a different source:

The knowledge of truth as such is wonderful, but it is so little capable of acting as a guide that it cannot prove even the justification and the value of the aspiration towards that very knowledge of truth. Here we face, therefore, the limits of the purely rational conception of our existence.[9]

Einstein does not, however, believe reason to be useless in our ethical pursuits, for he held that reason must be utilized to determine the proper means of application of ethical principles.

I know that it is a hopeless undertaking to debate about fundamental value judgments. For instance if someone approves, as a goal, the extirpation of the human race from the earth, one cannot refute such a viewpoint on rational grounds. But if there is agreement on certain goals and values, one can argue rationally about the means by which these objectives may be attained.[10]

Thus, while ethical judgments are not to find their

justification through logical deductions, Einstein was
vehement that reason not be dismissed in our ethical
activities:

> It might seem as if logical thinking were irrel-
> evant for ethics. Scientific statements of facts
> and relations, indeed, cannot produce ethical
> directives. However, ethical directives can be
> made rational and coherent by logical thinking and
> empirical knowledge. If we can agree on some fun-
> damental ethical propositions, then other ethical
> propositions can be derived from them, provided
> that the original premises are stated with suffi-
> cient precision. Such ethical premises play a sim-
> ilar role in ethics, to that played by axioms in
> mathematics.[11]

Thus Einstein placed ethical principles and goals on a level
analogous with the axioms of mathematics.
We are then permitted to introduce logical procedures of
deduction and derive ethical propositions from established
ethical premises. Such a characterization of ethical statements importantly separates Einstein from those, like the
logical positivists, who wish to label ethical claims meaningless.

> "We do not feel at all that it is meaningless to
> ask such questions as: 'Why should we not lie?'
> We feel that such questions are meaningful because
> in all discussions of this kind some ethical
> premises are tacitly taken for granted. We then
> feel satisfied when we succeed in tracing back the
> ethical directive in question to these basic
> premises."[12]

Let us attempt this tracing back to basic premises. The
ethical claim that we should never lie is based on the belief
that lying destroys our confidence and reliance in the statements of others. If we lack such confidence, social cooperation would ultimately become impossible. Such cooperation is
imperative, however, if human life is to be possible and
bearable. Thus, the directive "we should never lie" might be
traced back to the basic premises "human life shall be pre-
served" and "human pain and suffering should be lessened as
much as possible."[11] Such as example provides, if not the
rigor we might like, at least the flavor of what Einstein had
in mind.
Einstein believed that our conscious acts result from
very primitive desires and fears that are part of the very
nature of humankind. According to him "[w]e all try to
escape pain and death, while we seek what is pleasant. We
all are ruled in what we do by impulses; and these impulses
are so organized that our actions in general serve for our
self-preservation and that of the race." This is not to say,
however, that man's nature is purely of such an animalistic
character. "As social beings," Einstein noted, "we are moved
in the relations with our fellow beings by such feelings as
sympathy, pride, hate, need for power, pity and so on."
Thus, the root of all conscious action for Einstein is to be

found in the primary impulses which serve man's needs for self-preservation and societal interaction.[13]

This basic notion of primary human impulse led Einstein to several immediate conclusions concerning ethical actions. Rather than placing ethical doctrines in the context of religious dogma or faith, "[a] man's ethical behavior should be based effectually on sympathy, education and social ties; no religious basis is necessary."[14] However, while philosophers such as Plato found it necessary to claim that people will do what is best once they have been shown and understand what is best, Einstein was far more pessimistic concerning the power of rational persuasion for promoting ethic behavior: "I do not believe that the lack of moral and aesthetic values can be counter-balanced by purely intellectual effort." While education might promote the feelings of pity or love for proper moral behavior most people "are governed by passions among which hatred and shortsighted selfishness are dominant."[15]

Intelligence and rational thought, while not always determining the resultant action, were considered by Einstein to be an intermediate step between basic impulses and final action. Einstein's set of primary impulses could have been just as easily applied to the higher animals; man's rational character and use of linguistic symbols distinguishes him most prominently from animals. Thus, we must not allow individuals to surrender to the instincts of pleasure and pain for, as Einstein argued, this only leads to insecurity and fear. What rather should be encouraged is an increased attention paid to the emotions of love, pity, and friendship which, while admittedly weak, can be properly nurtured.[16] Such a conclusion brings us back to the basis of these ethical principles and to their apparent arbitrary nature. "For pure logic," Einstein held, "all axioms are arbitrary, including the axioms of ethics." But as is now clear, ethical axioms "are by no means arbitrary from a psychological and genetic point of view. They are derived from our inborn tendencies to avoid pain and annihilation, and from the accumulated emotional reaction of individuals to the behavior of their neighbors."[17] Thus, even the search itself for rational justification of ethical principles is not rationally justifiable but is a result of basic instincts and drives.

Rational attempts at metaphysical and ethical understanding are for Einstein, clearly limited by our conceptual and axiomatic choices. More generally, Einstein held that reason is limited by its own inability to justify the value of the aspiration to achieve knowledge or truth.[18] It is to this more general limitation that I wish to offer some concluding speculative thoughts.

What Einstein seems most importantly to have captured about rational investigations is that they are generated by the nonrational feelings. Most important among these feelings, for Einstein, were wonder and awe.[19] From our amazement and wonderment at the world around us there is generated the need to comprehend and the asking of the question why. What Einstein so ingeniously saw, however, was that while these explanations and systems are themselves limited by freely chosen axioms and concepts, comprehensibility and knowledge can nonetheless be achieved. It is the fact that we come to comprehend and understand that then produces in us

86 Richard Fleming

an even more sophisticated sense of wonder. The most incom-
prehensible thing about the universe, Einstein remarked, is
that it is comprehensible.[20] This more sophisticated set of
feelings then generates anew the need to comprehend, the asking of the question why of those particular concepts or
axioms. Why just these and not others? So the process continues ad infinitum with every new rational explanation
bringing with it new concepts and feelings which demand
further explanation. Rather, therefore, than viewing rational disciplines as attempts to provide final and complete
answers to the peculiar questions of those disciplines, it
seems more correct to stress the limitations placed on those
inquiries by their conceptual choices, to emphasize not the
resulting end or answer that might be forthcoming, but to recognize instead the importance of the constant striving that
is a part of, not only some particular rational investigation, but the very heart of rationality itself.

Notes

1.   Albert Einstein, Max Born, and Hedwig Born, The
Born-Einstein Letters (New York: Walker and Co., 1971), p.
163.
2.   Albert Einstein, "Remarks on Bertrand Russell's
Theory of Knowledge," in Ideas and Opinions [henceforth, IO]
(New York: Dell Publishing Co., 1973), p. 34.
3. Einstein, "The Problem of Space, Ether and the Field
in Physics," in The World As I See It [henceforth TWS] (New
York: Covici, Friede Pub., 1934), p. 84.
4.   Einstein, "Physics and Reality," in Out of My Later
Years [henceforth, OLY] (Totowa, New Jersey: Littlefield,
Adams & Co., 1967), p. 63; "Remarks on Bertrand Russell's
Theory of Knowledge," IO, p. 33; "On the Methods of Theoretical Physics,," TWS, pp. 35 - 36.
5.   Einstein, "Clerk Maxwell's Influence on the Evolution of the Idea of Physical Reality," TWS, p. 60.
6.   Albert Einstein and Leopold Infeld, The Evolution of
Physics (New York: Simon and Schuster, 1938), p. 125.
7.   Albert Einstein, "Reply to Criticisms," in Albert
Einstein: Philosopher-Scientist, vol. 2 [henceforth, EPS],
ed. P. A. Schilpp (New York: Harper and Row, 1959), p. 667.
8.   Einstein, "Reply to Criticisms," EPS, p. 683.
9.   Einstein, "Science and Religion," OLY, p. 26.
10.   Einstein, "On Freedom," OLY, p. 18.
11.   Einstein, "The Laws of Science and the Laws of
Ethics," OLY, pp. 110 - 11.
12.   Ibid.
12.   Einstein, "Morals and Emotions," OLY, p. 20.
14.   Einstein, "Religion and Science," TWS, p. 266.
15.   Otto Nathan and Heinz Nordon, eds., Einstein on
Peace (New York: Schocken Books, 1968), p. 556.
16.   Einstein, "Morals and Emotions," OLY, pp. 21 - 22.
17.   Einstein, "The Laws of Science and the Laws of
Ethics," OLY, p. 111.
18.   Einstein, "Science and Religion," OLY, p. 26.
19.   See for instance, Einstein, "The World As I See
It," TWS, p. 242; "Religion and Science," TWS, p. 265; "Autobiographical Notes," EPS, vol. 1, p. 9.
20.   Einstein, "Physics and Reality," OLY, pp. 60 - 61.

10.

Ontological Relativity: A Metaphysical Critique of Einstein's Thought

ASHOK K. GANGADEAN

Einstein's thought, when placed in the context of the logic of interpretation, may be seen as part of a long tradition of ontological thought where radical revolutions in making sense of things have been the rule rather than the exception. I shall attempt to show that Einstein's equation $(E=mc^2)$ is an <u>ontological</u> equation that has categorial significance and redefines certain basic categories in his ontology. We shall see that the history of ontological revolutions, in both the Indian and Western traditions, reveals that there are common categorical principles and patterns of thought which govern the formation and transformation of ontological paradigms.

Thus, by placing Einstein's thought within the tradition of ontological revolutions we are able to detect what is both new and old in his ontology. I shall attempt to show that in developing relativity theory he opened the door to "ontological relativity" which, for Western scientific thought, is a radically new rationality for making sense of the world. Einstein opened the way to overcoming ontological absolutism, but he nevertheless remained committed to an absolutist ontological mentality. This created a deep tension in his thought which, apparently, he was not able to overcome in his struggle for a unified ontological paradigm. The revolution which he started is now being played out and remains unfinished. A critical transformation in the unfinished revolution is to find the ontological equation which connects consciousness as a form of energy with physical energy as this was clarified in Einstein's ontology. The discovery of such a connection would open the way to accounting for a vast range of phenomena which are not intelligible: for example, psychokinesis, human volition as a form of psychokinesis, so-called psychosomatic diseases, and the apparent dual nature of light. At the end of this chapter I shall suggest a possible ontological transformation which reveals the relation between consciousness-energy and physical energy.

Ontology is the science of being and thereby the science of sense. It investigates the basic categories which configure to define reality. Of course, this does not mean that there is a stock of basic categories which have independent and autonomous meaning. Rather, the meaning of a category is <u>itself</u> conditioned and defined by the configuration or

structure in which it is manifested. Categories are dialect-
ically and organically defined by their particular configuration. Such a configuration is called an <u>ontological</u>
<u>paradigm</u>. But it must be stressed that such a paradigm is
itself the origin of meaning and makes experience possible.

The purported completeness of a world view is not com-
promised by the fact that ontological hermeneutics discerns
different possible worlds. Both the Hindu world and the
Christian world present themselves as being universal, complete, and exhaustive of reality and forms of human experience. Yet what counts as possible phenomena or facts in one
world is not recognizable or intelligible in the other. In
the Hindu world, for example, yoga is a rigorous science
which naturally flows out of that world view. In the
Christian world science takes a different form. It is naive
to assume that there is one empirical science for all possible worlds.

Now, an ontological paradigm is not what is commonly
called a conceptual framework or theoretical paradigm in
natural science. One way to see this is to notice that a
category is <u>not</u> an ordinary concept, and that ontology investigates <u>sense</u> not factual truth. In categorial logic a category is defined as a fundamental or primary formal concept
which exhausts some domain of the universe. For example, the
term "red" specifies a property. Any such term has a logical
contrary which, when taken together, define what is called an
"absolute term" (in this case, "being a colored thing").
Thus, "red" and its contrary "un-red" together exhaust a
range or <u>category</u> of reality. If we indicate any absolute
term within slashes, then "red-or-un-red" is written /red/.
Let us, then, reserve the term "category" for such feature
terms and use "concept" for property terms.

Moreover, a new ontological paradigm redefines radically
the range of possible phenomena. What emerges as intelligible phenomena in Einstein's world, for example, are not
possible phenomena within Newton's ontology. Yet, it is
simply misguided to maintain that certain phenomena which are
intelligible and commonplace in one world are "not real."
Verification claims across possible worlds are problematic.
The cross-ontological verification claims of a particular
scientific ontology are reflections of the categorial limits
of that world rather than legitimate truth-claims about the
alien world.

In ontological revolutions a given ontologist argues
that his new paradigm is preferable since it makes more sense
of things. Yet, the logical principles which are valid for
facts <u>within</u> a given world do not apply to the ontological
"affirmation" of a world taken as a whole. Different worlds
do not stand in the oppositional relation which propositions
and facts do within a given world. The oppositional relation
between propositions within a world is bound in an univocal
relation of shared sense. There must be univocity of sense
between a proposition and its opposite. But this is precisely what is formally precluded between different worlds. To
"affirm" a world as a whole does not entail the "denial" of
other worlds. For it is ontological commitment which gives
rise to meaning itself, and thus to the possibility of truth
and falsity. This is why it is appropriate to speak in religious conversions (ontological revolutions) of being born

again, taking on a new identity, and finding meaning in life.
     Moreover, it must be seen that rational faith is the
origin of scientific life. Within such a commitment the
scientist may seek truth. But the form of life itself is not
open to rational testing and validation. In this respect the
scientist is just as much lodged in faith as the religious
person. Indeed, the life of Einstein is exemplary in this
respect; for he made it clear that he regarded his life in
science as being the same as his religious life.
     Whatever the <u>differences</u> may be between religious con-
versions and scientific revolutions, there nevertheless remains a common logic of ontological transformations. There
is a characteristic pattern in an ontological revolution, in
the transformation of sense from one system to another.
First, when the revolutionary statements are first made they
appear to be <u>category mistakes</u> with respect to the generally
accepted language. Einstein's observation that light bends
in the gravitational field of the sun sounds like a category
mistake in the Newtonian language. For it presupposes a connection between the categories of /matter/ and /energy/ which
does not obtain in Newton's world. Second, the transformation to a new system of sense, a new categorial structure, is
aided through treating the new strange utterances as metaphors. Metaphors are the sense vehicles of transformation
between paradigms. They may be regarded in this function as
"creative category mistakes." Third, as the new categorial
paradigm is entered the utterances are no longer strange, but
become literal and univocal in the new world: of course
light bends, of course mass is a form of energy, and of
course space and time are relative. The absolutist mentality
makes the mistake of insisting that, like incompatible propositions or theories within a given world, ontological paradigms too are incompatible in the same way: one is true and
more valid than the rest. This mentality treats alternative
worlds or realities as if they were competing theories. It
pretends that although there may be different world views,
there is nevertheless one common reality with "objective"
preontological "facts." But this amounts to holding that
there is really one and only one world. This begs the question of there being genuinely alternative ontologies.
     Ontological relativity takes seriously that there are
<u>different possible worlds</u>. This means that different system-
atic forms of intelligibility are simultaneously valid.
Ontological relativity means a relativity of /truth/.
Natural thought and language, natural reason, are seen to be
ontologically relative, and this is reflected in the relativity and indeterminancy of the sense of any conceptor term.
Any term in natural language remains formally open to rein-
terpretation relative to some other ontology.
     Ontological relativity must not be confused with rel-
ativism, with the position that truth is subjective and a
matter of mere opinion. Nor does Einstein's relativity
theory entail relativism in this form. His relativity did
entail the rejection of a form of absolutism concerning space
and time in Newton's ontology. It is commonly known that the
discovery of alternative coherent and consistent geometries
opened the way for the discovery of relativity of space.
Similarly, the recognition of the fact of alternative
coherent and consistent ontologies or "geometries of being"

requires the transformation to ontological relativity. Reality can no longer be identified with any one ontology, but neither can it be postulated apart from any ontology. Ontological relativity, in fact, leads to the true absolute, while ontological absolutism degenerates into a form of relativism.

But ontological relativity is an ancient discovery. It goes back as far as the teaching of the great Buddhist dialectician Nagarjuna (the second - century founder of the Madhyamika tradition). Nagarjuna, in explicating the teachings of the Buddha, demonstrated that all views are <u>sunya</u> or empty. In effect, he attempted to show that any form of ontological absolutism leads to incoherence and unintelligibility. The common commitment of absolutism is that reality is ontologically determinate and self-existent, that there is an ultimate substantial referent (svabhava). His dialectic was designed to reveal that any attempt to fix reality, to grasp the true essence of things, must be self-defeating and inconsistent. Nagarjuna attempted to demonstrate that this absolutist mentality is the source of all suffering and mental disturbance. He teaches that one enters true consciousness when the rationality of ontological relativity is realized.[1]

Nagarjuna, for example, showed that the structure of rationality is bipolar; no matter what ontological paradigm is considered, or whatever entity is scrutinized, bipolarity is formally present. The Hindu paradigm (the atma ontology) showed one side of the bipolarity in developing to its limits the monistic substantial pole, while the early Buddhist ontology (the anatma ontology of dharmas) developed to its limits the atomistic nonsubstantial pole. Both mentalities were absolutist according to Nagarjuna. Each pole is incomplete when taken independently, but taken together they yield ontological relativity and transcategorial awareness. Each pole of dipolar rationality appears incompatible and inconsistent with the other in absolutist logic, yet they complement each other in the logic of relativity. It should not be surprising then, that transcategorial (bipolar) entities will inevitably emerge within any given ontological paradigm. It was to be expected that theoretical physics would eventually encounter ontological relativity, relativity of sense and essence, indeterminacy and inherent ambiguity (as in Werner Heisenberg's thought), transcategorial entities, and bipolar paradigms (particle theory, which is atomic, and field theory, which is monistic and continuous).

Therefore, the absolutist problematic concerning the nature of ontological commitment dissolves. The question, "Which among possible worlds is the true one?" is a transcategorial mistake. For the absolutist rationality is <u>itself</u> in question, and far from explaining the nature of ontological revolutions, it makes such radical transformation unintelligible. It is here that we find the key to <u>authentic</u> ontological commitment. Ontological commitments which flow from the hermeneutic of the absolutist rationality are ontologically bound, incomplete, and incoherent. To say "I am Christian," or "I am Hindu," or "I am Buddhist," or "I believe in science" in an absolutist confession of faith is unliberated, bad faith. But false consciousness is overcome

and liberated in the mentality of ontological relativity: to
say "I am Christian," or "science is true" in the hermeneutic
and rationality of ontological relativity is good faith.
This difference in hermeneutical attitude makes all the
difference in the world. For interpretation is not merely an
epistemic state of mind but rather constitutes our being-in-
the-world. When rationality itself matures beyond the absolutist hermeneutic and realizes its fruition in ontological
relativity, consciousness is liberated and is no longer
alienated from what is /true/.

Although Einstein's discoveries in relativity theory
opened the way to general ontological relativity, he nevertheless remained committed in the absolutist mentality to a
particular ontological paradigm. First, the inherent limits
of the physicalist paradigm created an alienation from a vast
range of phenomena which involved the presence of consciousness as a form of /energy/. Second, the absolutist mentality
itself alienated scientific interpretation from the inevitable discovery of transcategorial entities which could not be
accommodated in any one paradigm. And third, the absolutist
hermeneutic repressed the realization of the essential bipolar form of rationality which alone can truly reveal a unified paradigm which can accommodate the apparently inconsistent phenomena which are at once both atomic, and anatomic.

The suggestion that reality is ultimately contradictory
and inconsistent violates our rational sense. Einstein
apparently found the dualism at the heart of physics to be
unacceptable and strove to find the deeper unity which, he
assumed, must be there. Although he was creative in both the
wave and particle theoretical models, his deeper commitment
was to field phenomena. For quantum theory revealed nature
to be discontinuous, indeterminate, and probabilistic, while
field theory stressed continuity, uniformity, and rational
determinacy, which, for Einstein, promised some hope of ultimate unification. So Einstein's sympathies led him away from
quantum theory as the unified foundation of physics.

Einstein's relentless quest for a truly unified paradigm
of interpretation in the face of the dualism in theoretical
physics was at once a mark of his creative genius and
integrity as well as of his ontological limitations. It was
the mark of his genius because he was creative and innovative
on both sides of the dualism, in both wave and particle
theory. It was a sign of his character and integrity because
despite his deep need for theoretical unity, he was unwilling
to compromise and "find it" in one or the other side of the
dualism. It was a signal of his ontological limitations
because he remains so deeply committed to absolutist rationality and physicalist ontology that he left himself no option
but to seek the unity of interpretation and overcome the
dualism at the foundations of physics by reducing one side of
the dualism to the other. His program was to show that a
unified field theory would unite all phenomena, and this
implies that the laws of quantum theory would find their
deeper meaning in the laws of the field.

That Einstein's conceptual revisions involve a categorial revolution may be seen in the fact that new sense relations become possible. In Einstein's ontological language
it makes sense to speak of light (a form of energy) bending
in a gravitational field, and having mass. It does not make

sense to speak in this way in Newton's categorial paradigm: it would be a category mistake, neither true nor false. Moreover, all of the categories of Einstein's new language are mutually related in sense; each shapes the sense of the other. Space is intimately bound with time, time is intimately bound with the velocity of energy, energy is bound with mass or matter, matter is determined by velocity, which in turn is delimited by the absolute speed of light, but space itself is shaped by the velocity of masses in the gravitational field which determine the movement of time, and so on. In Einstein's language we find that space and time form a categorial continuum, and in effect we now have one category: /space-time/. It was through this more primitive category that Einstein was able to uncover a deeper unity connecting "Newtonian phenomena" with the field phenomena of Maxwell's electromagnetic theory. Yet strictly speaking, it is a <u>paradigm crossing</u> (global category mistake) to say that the phenomena of Newton's world are encompassed within Einstein's universe. "Newtonian phenomena" have been redefined into <u>field</u> phenomena, and <u>gravitation</u> itself has become a radically new concept. So the new unity, simplicity, and generality achieved in Einstein's categorial language come about through the reconstitution of physics itself into relativistic mechanics.

For example, "time" in Einstein's world means <u>measured time</u>, a referential time which is necessarily relativized to an moving event-system. Such measured time is variable since as a moving clock approaches the speed of light its rhythm slows down, and if it could reach the maximum possible velocity, time would stop altogether. It should be stressed, however, that time slows down only with respect to a specific event-system, not to the cosmos as a whole.

Einstein's language appears to be coherent and consistent as long as we restrict attention to a specific event-system. But the moment we attempt to become cosmic and universalize for all possible systems <u>at the same time</u>, that is, to get a unified and coherent ontological grasp of the universe <u>as a whole</u>, a deep incoherence emerges. For, on the one hand, ontological cosmology requires that the universe be <u>one</u> in some sense. But, on the other hand, if time is relativized in the way Einstein suggests, and there is no cosmic absolute /time/, then existence itself becomes pluralized and atomized into different discrete and independent universes. So Einstein cannot coherently maintain at the same time that there is only relative time <u>and</u> there is one universe, since, insofar as we can conceive of a /universe/ as a whole absolute /time/, /energy/, and /existence/ are presupposed.

The only way to achieve ontological coherence is to recognize that metatime or cosmic absolute /time/ is presupposed. Ontological /time/ neither moves nor fails to move. Rather, all /movement/ presupposes /time/ as a transcendental condition. It is /time/ which substantiates and unifies the existential field in event-ontology. Einstein's <u>moving time</u> (measured, referential time) presupposes cosmic nonmoving /time/. Measured time, clock time, is one conceptual interpretation of the absolute category of /time/. It is cosmic /time/ which unites the range of discrete event-systems into one /universe/.

Thus, the category of /ether/ introduced into the

Newtonian ontology may not have been a bad idea after all. Properly understood, /ether/ is the uniform substance or medium of /space/ which neither moves nor rests. It is the unifying principle or category required by the Newtonian ontology to accommodate the field phenomena of electromagnetism. Of course, the /ether/ as the universal transcendental condition of all wave-motions could not be detected by experimental observation, for it was a metaphysical condition of the possibility of coherent experience within Newton's mechanical universe. And, if the preceding line of thought is valid, Einstein's ontology too requires its own "/ether/" in the form of cosmic /time/. Once we recognize the ambiguity between measured clock time and cosmic /time/ we are not tempted to expect that we could detect /time/ empirically, nor is it an embarrassment for physical ontology. But this does not mean that it is not real.

From the perspective of ontological hermeneutics the most remarkable feature of Einstein's thought is that he was creative in the two formal complementary ontological languages. Although he never gave up the hope of finding a unified field theory, he nevertheless contributed to the foundations of quantum theory, which essentially involves atomistic thought. But it should be clear that the two paradigms are not contrary opposites, but contradictory opposites. That is, each paradigm formally excludes the other.

We saw earlier that conceptual differences may be reconciled in the more primitive unity of a category. Each category has an internal logical structure of contrariety: if /P/ is a category, its structure consists of the contraries "P" and "un-P," where each is opposed and yet dependent in meaning upon the other. Contraries are mutually inclusive and included within the category they constitute. In this respect, the logic of contrariety has governed and dictated many critical "discoveries" in contemporary physics. Where "matter" is taken as a concept it is natural to "discover" antimatter as its contrary, each opposed, yet together comprising the category of /matter/.

However, the opposition between atomism and anatomism is of a different nature — they are contradictories rather than contraries. There can be no primitive unifying category which can hold them together in contrary opposition. They are transcategorial opposites, while contraries are intracategorial opposites. There is no conceivable transformation from atomic phenomena to anatomic phenomena, for they are divided by radical systematic ambiguity. To suggest that a given phenomenon or entity was at once both atomic and anatomic would be to propose a contradiction — to violate the logical principle of noncontradiction and its corollary principle of identity. Any attempt to combine atomism and anatomism in the concept leads to the incoherence of a paradigm crossing. On the other hand, the attempt to synthesize the two by incorporating one into the other falls into the incoherence of reductionism. Ontological hermeneutics reveals the depth of this problem in the form of a dilemma which challenges absolutist rationality. This is Einstein's transcategorial problem — reality can be one and univocal in the face of bipolarity only with a radically new form of rationality — only with transcategorial unity. The transcategorial predicament of physics is just a more recent

recurrence of an ancient ontological problematic. It was indicated earlier that Einstein shared with Plato a commitment
to absolutist rationality. But the genius of Plato did not
hold back from making the full move to transcategorial
being. Plato attempted to preserve absolutist reason by
dualizing his ontology into the world of change and the world
of nonchanging forms or essences. This was his strategy to
save rationality from the threat of incoherence and contradictions in the ontological discoveries of Parmenides and
Heraclitus. How could /being/ be both at /rest/ and in
/motion/ at the same time? The answer was that there were
two worlds. But Plato's quest for unity beyond this duality
required him to make the transcategorial move to /goodness/,
which was beyond /being/ and /truth/. /Goodness/ was the
ultimate form of forms which was the source of rational
light. In this ultimate transformation Plato made the right
move to transcategorial rationality and unity. In this respect Plato recognized special ontological relativity.

Nagarjuna also moved to general ontological relativity.
This is achieved only when the formal bipolarity of atomism
and anatomism has been explicitly developed. This is what
Nagarjuna did when he shows the antomism of the Hindu atma
ontology, and the atomism of the Buddhist anatma paradigm.
The genius of Nagarjuna was to have seen that only a radical
revolution in reason itself, beyond absolutism and essential-
ism, could achieve the rational unity and coherence of experience.

But now another prejudice of the ontology of physics
must be mentioned. It is the physicalism of physics — the
commitment that /energy/ must be physical in nature. It is
only now beginning to become clear that consciousness cannot
be coherently excluded from the categorial structure of
physics. Quantum physics has recognized that the presence of
consciousness must be included in any complete account of
physical phenomena or reality. And, as before, the perennial
danger is that of reducing the category of /consciousness/ to
the category of /matter/. Physical ontology becomes a
generic and universal holistic ontology when it "discovers"
that consciousness is a form of /energy/, of cosmic energy.

As the science of /energy/, holistic physics will then
be in a position to account for a vast range of everyday
phenomena which now completely escapes its current categorial
structures. For example, every instance of awareness or
thought moving the human body, as in the act of volition, is
an example of psychokinesis, of the mind moving matter. But
a most dramatic example in which the energy of consciousness
transforms the world is to be found in the thought of
Einstein. His ontological and conceptual transformation of
human understanding changed the physical world in a radical
way, indeed, it opened up a new universe. All such transformations between thought or /consciousness/ as a form of
/energy/ and the physical world of physical energy call out
for rational explanation in holistic physics.

Notes

1. These points can only be sketched briefly here. For
a more systematic and developed discussion of Nagarjuna's
teaching see the following papers: A. K. Gangadean, "Formal

Ontology and the Dialectical Transformation of Consciousness," Philosophy East and West 29, no. 1 (January, 1979). "Nagarjuna, Aristotle and Frege on the Nature of Thought" in Buddhist and Western Philosophy, ed. Nathan Katz (Sterling Publishers, New Delhi, 1980); "Comparative Ontology: Relative and Absolute Truth", Philosophy East and West 30, p. 466 - 480 (1980).

Part V

History and Philosophy of Science

11.

Relativity Before Einstein: Leo Hebraeus and Giambattista Vico

WILLIAM MELCZER

The positivistic, crudely mechanistic, and essentially rationalistic science and philosophy of the late eighteenth and nineteenth centuries, against which Einstein's theory of relativity constituted a speculative reaction, may be compared with at least two previous moments in European intellectual inquiry: the scholastic Aristotelianism of the late Middle Ages and the early Renaissance and, in particular, early quattrocento Albertian-Aristotelian civic-humanistic thought; and the Cartesian rationalism of the seventeenth century with its rigid syllogistic apparatus, schematized geometrical methodology, and exacting and immutable frame of reference. Most importantly, each of these two moments of essentially rationalistic inquiry propounded a <u>finito</u> conception of the universe as well as a <u>finito</u> conception of the categories and the forces acting and interacting in that universe.

These two moments of rationalistic inquiry were followed by intellectual and philosophic reactions oriented towards a breakdown of the absolute, <u>finito</u>, rationalistic categorizations: Cusanus and the subsequent Florentine Neoplatonism, while rejecting the rigid Aristotelian categorizations, establish a syncretist philosophy aimed at a <u>non-finito</u> universe; and early eighteenth-century Italian humanistic thought subjects Cartesianism to a devastating critique.[1]

By the end of the fifteenth century, new intellectual forces make their appearance on the European scene: Cusanus, Ficino, Pico della Mirandola, and Leo Hebraeus work, though not always consistently, towards the dismantling of the fixity of the Aristotelian categories. At the same time, they steadily push their theoretical speculation towards the establishment of a purposeful mobility of the universal categories. Within these, a place of particular conceptual eminence is assumed by the human category, more precisely, by the human soul: hence the spectacular development of the concept of the dignity of man. The mobility of the categories, however, constitutes but one of the abiding interests of this speculative current; some of their other absorbing preoccupations are the conception of a <u>non-finito</u> universe both in terms of its parameters and in terms of the forces governing it, and the establishment of a syncretist Neoplatonic synthesis of Plato, Aristotle, Plotinus, and various theological and philosophical constructs.

In 1535 Leo Hebraeus' Dialoghi d'amore was published in Rome.[2] The Neoplatonism of the work is strongly Plotinian. Its main tenet consists in the assumption that love is the preeminent bond of affinity in a universe that is neither theocentric, as Dante's was, nor anthropomorphic, as that of the Albertian-Aristotelian quattrocento, but which is forever dynamically oscillating between God and man, or better, between Creator and the created universe. Governed by the forces of universal affinity, categories lose their fixity and transmute into each other. Everything turns, or, at least, may turn, into everything else.

In addition, a remarkably dynamic conception governs the thought of Leo Hebraeus in his assigning a pivotal position to love within the metaphysics of the Dialoghi. Love is conceived of, not so much as a centrally located power or essence, but rather as an omnipresent, ubiquitous, agglutinative, and, on the whole, cementing bond that holds together, through never ceasing dynamic motion, the Creator and the created world. Love is, in fact, the cause for the original creation and also for the subsequent cohesive togetherness of the universe:

> The world and things have being insofar as they
> are all united and linked with all its things as
> the limbs of an individual.... and since nothing
> makes the universe united with all its various
> things if not love, it follows that love is the
> cause of the being of the world and all its
> things.[3]

The concept of love as an universal principle Leo derives directly from Eryximachus' discourse in Plato's Symposium, probably in Ficino's translation. What is remarkable in this case is the use our philosopher makes of this theory, by converting the universal principle of love into the causal condition for the very existence of the world. Somewhat earlier, in the same dialogue, he writes: "Not only beatitude would be missing if there was no love, but neither the world would have being [that is to say, existence], nor would anything be in it, if it were not because of love."[4]

In these and other remarkable passages Leo Hebraeus comes to suggest, without stating it explicitly, that love is not merely a condition of beings, but the condition of being itself, and, actually, that love is being itself. There is no doubt that, though couched in an esoteric Neoplatonic erotic terminology, our philosopher gravitates here towards the notion that categories of beings are ultimately relations and, truly, that things are ultimately relations. With such an eminently dynamic concept Leo at once becomes the forerunner of the nature philosophers of the following generations and, perhaps even more importantly, of modern physics.

In the Dialoghi d'amore the Aristotelian law of causality, in which cause is necessarily anterior to effect, is transcended. Leo Hebraeus develops a metaphysics in which, though knowledge is a condition of love, it becomes also a consequence of love. Like the concept of causality, Aristotelian time is also transcended.

But Aristotelian causality is transcended in a more sub-

tle way too. Our philosopher establishes unmistakably the
preeminence of the paternal love of participation over the
filial love of union. Now, the filial love of union is indeed founded on a straightforward cause-effect relationship:
it is the love of a higher being by a lower one in which the
lower being's motivation consists in the acquisition of
beauty he does not possess. But paternal love is different.
Superior beings find in their inferiors no beauty they would
not possess, writes Leo Hebraeus quite surprisingly, but the
desire of beauty not yet possessed is not at all the cause of
their love. Superior beings have an innate desire to impart
to their inferiors their own beauty, and by such imparting of
their beauty they themselves become beautified.[5]

Thus, without rejecting the Platonic theory of the pursuit of the beautiful as one of the legitimate objects of
love, Leo Hebraeus complements this concept with a second
kind of love that is based upon giving rather than upon re-
ceiving. Descending paternal love, then possesses a gratui-
tous quality. While ascending love is a desire for beauty
one does not possess, descending love is a bestowing of
beauty that others do not possess. In such a scheme of
things though there is a built-in paradox — a Christian
paradox, so to speak: by giving beauty to others, oneself
too is beautified. Such a paradoxical beautifying of oneself
is only conceivable in a metaphysical organization that leans
towards an immanentistic conception of the universe.

We must turn now, in the briefest terms, to Vico.
Giambattista Vico's De nostri temporis studiorum ratione
(1708) signals, in the evolution of its author's thought, the
passage from a rhetorical to a philosophical attitude.[6] At
the same time, the work establishes the propaedeutic orientation of Vico in relation to the contemporary Cartesianism. A
reevaluation of a number of Cartesian elements takes place
here: ingenium is pitted against reason; topica, against
Cartesian criticism; invention, against demonstration;
eloquence, against syllogism. Rigidity, the lack of human
warmth, and arid schematization are the negative connotations
attached to Cartesian rationalism and to its method of
inquiry in this onslaught of newly developed humanistic
critique. In such an educational context, topos (that is to
say, argumentorum inventio) precedes and conditions all
rational critique. The general propaedeutical function of
the topos, recognized already by Aristotle, becomes for Vico
a necessary propaedeutical function determining the prera-
tional elements of the mind and conditioning thus the ensuing
rational critique. Thus, since the topos is variable, different in each case, so the ensuing ratio will be variable
and different too. This is tantamount to saying that the rational faculties of men are in themselves variables, and
hence, that ratiocination itself loses its connotation of a
reliable, base-line, common denominator for the pursuit of
cognizance.

The classical, aesthetic, and humanistic interests conceived in their historical relativity are further elaborated
upon in the metaphysical portion of the De antiquissima
Italorum sapientia (1710) and, in particular, in its gnosio-
logical doctrine of the verum ipsum factum. According to
this remarkably encompassing theory — essentially, an
immanentistic conception applied to epistemology, a concep-

tion in which the cognizant subject tends to become one with the _cognitum_, actually, the former creates the latter — the knowledge of a thing consists ultimately in the historical reconstruction of its being.

Suffice it to say here that, applied to the _devenir humaine_ (the social concerns of Vico), what this really means is that there is no possibility for an absolute historical knowledge, because no historical reality can indeed be totally reconstructed and also because no two reconstructions are identical. Such a deep-seated historical relativity pulls the rug from under the feet of the sacrosanct historical fact standing on its own in splendid isolation and waiting for the equally armed cognizant minds to fetch it.

Leo Hebraeus' theory of metaphysical relativity by which the Aristotelian law of causality is transcended and by which categories become relations, as well as Giambattista Vico's theory of psychological and historical relativity by which _topos_ is made to precede and to condition all rational inquiry, and by which the knowledge of an event is determined each time anew, and each time variously, by its historical reconstruction, constitute early, parallel, and organically connected stages of relativistic thought heralding from the distance the scientific relativity of Einstein.

The lesson to be derived from the above speculative rapprochement is that the thinking modalities of the mind operate quite independently from the objective areas of concern, that is to say, the disciplines — philosophy, theology, historiography, or natural science. Without claiming total independence between the intellecting mind and the object of inquiry — the cognizant subject and the _cognitum_ in the case of Vico — the above instances seem to point at least a substantial degree of auto-sufficiency of the mind which reverts, at determined historical junctures, to a _non-finito_ and relativistic conception of the universe.

Notes

1.   The question of the _non-finito_ is treated exhaustively by M. de Gandillac, _La philosophie de Nicolas de Cuse_, (Paris, Aubier, Éditions Montaigne, 1941).

2.   The work has been little studied; the author is relatively little known. See J. de Carvalho, _Leao Hebreu, Filosofo_, (Coimbra: Imprensa da Universidade 1918); F. Zimmels, _Leo Hebraeus, ein judischer Philosoph der Renaissance_, (Breslau: W. Koebner, 1886; E. Solmi, _Benedetto Spinoza e Leone Ebreo_, (Modena: Coi tipi di G. T. Vincenzi e Nipoti, 1903; C. Gebhardt, "Einleitung zu Leone Ebreo," in Leone Ebreo, _Dialoghi d'amore_ (Heidelberg: C. Winter, 1924); H. Pflaum, _Die Idee der Liebe / Leone Ebreo_ (Tubingen: J. C. B. Mohr, 1926; G. Saitta, "La filosofia de Leone Ebreo," in _Storici antichi e moderni_ (Venice, 1928); S. Damiens, _Amour et intellect chez Léon l'Hébreu_, (Toulouse: E. Privat, 1971); and my "Platonisme et Aristotélisme dans la pensée de Léon l'Hébreu," in _Platon et Aristote a la Renaissance_ (Paris, 1976 [Colloquium CESR, Tours, 1973]).

3.   Leone Ebreo, _Dialoghi d'amore_, S. Caramella editor, (Bari: G. Laterza & figli, 1929), p. 165.

4.   Ibid. p. 164.

5.   See my Platonisme et Aristotélisme dans la pensée de

Léon l'Hébreu."

6.   For the present discussion of Vico, see my "The
Historical Humanism of Vico and Present Day Technological
Science," in Vico-Venezia, ed. G. Tagliacozzo (place and date
yet unknown), as well as the following: E. Grassi, "Critical
Philosophy or Topical Philosophy?   Meditations on the De
nostri temporis studiorum ratione," in G. Tagliacozzo, ed.
Giambattista Vico; an International Symposium (Baltimore:
Johns Hopkins Press, 1969); and Y. Balaval, "Vico and Anti-
Cartesianism," in the same.

12.

Einstein, Extensionality, and the Principle of Relativity

DENNIS A. ROHATYN and PATRICK J. HURLEY

In this chapter, we discuss how one of Einstein's famous thought experiments would be viewed by a proponent of extensionalism and defend Einstein's own interpretation as being at least as plausible as the extensionalist alternative. Extensionalism can be described in terms of the adoption of the following principles:

1. The notion of a single reality is meaningless. By "meaningless" is meant "unverifiable, untestable in principle." Consequently, so is the notion of a single or all-embracing account of reality. Such notions do not advance or contribute to our knowledge.

2. What we find in the world varies in direct proportion to our behavioral set, as it is conditioned by language, culture, environment, and biology. The result of this conditioning is expressed, sometimes only implicitly, in our choice of a framework for the explanation of phenomena, and in our adoption of axioms and postulates used to give order to empirical or mathematical data.

3. Conflicts between differing or competing accounts of phenomena cannot be settled by appealing to reality, since (as (1) makes clear) such appeals are vacuous, and because (as (2) makes clear) the conflicts reflect different outlooks on the world dictated, at least in part, by different selections of assumptions. There is no neutral ground upon which to decide the relative merits of one axiom-system or another, since any standpoint that is available will amount to the choice of another framework of assumptions altogether.

4. Therefore, the choice between rival hypotheses in science is always a matter of arbitrary preference. Although criteria such as elegance, simplicity, and parsimony can be drawn upon, these are neither exact nor ontological in scope; they are at best appeals to our aesthetic or moral sense, which is notoriously vague as well as variable. In the last analysis, a choice reflects nothing more than the preoccupation or, if you like, the bias of the chooser.

Let us begin by examining one of Einstein's most celebrated thought experiments. In one of these, a cubic room containing an observer is moving rapidly to the right with respect to a second observer located outside the room.[1] In the center of the room a light is momentarily switched on, and the rays of light travel outward in all directions toward the walls. For the observer inside the room these rays reach all the walls simultaneously; but for the outside observer the left-hand wall seems to be approaching the oncoming ray, while the right-hand wall seems to be escaping from the ray. For the outside observer, therefore, the ray traveling to the left reaches that wall before the ray going to the right reaches that one.

In this experiment it is important to note that the two observers seem quite literally to see different things. This leads one to ask what the truth in these situations is, that is, what actually has happened? If each observer is permitted to define himself as constituting a laboratory, inertial framework, reference or coordinate system, it would appear that they are each entitled to describe their separate realities, a fantastic notion which seems more appropriate to science fiction than to science.

From the extensionalist's point of view, the implications are clear. "Reality" is either a meaningless word, or else becomes synonymous with "the operations that can be performed at a given time and in a given region of space," that is, a local (inertial) frame. Less technically, "reality" is context-relative, that is, relative to an observer. Since reality is relativized in this way, seemingly in keeping with the spirit of Einstein's theory, it follows that reality changes from one context to another, inasmuch as perceptions do, relative to given coordinate systems. This seems a victory for Heracliteanism, but the extensionalist eschews flux as much as he does permanence; he simply confines himself to remarking that as coordinate systems change, so does what we say about the world and, consequently, so does the world. If this seems baffling or counterintuitive, then the extensionalist will remind us that it is our own fault, for choosing to retain the word "reality" with all of its traditional and, therefore, misleading associations. Hence, science cannot concern itself with unanswerable questions such as "what is real?" except insofar as they can be translated, without loss of content, into questions such as "what is measurable, here and now?" The unanswerability stems not from human limitations, nor from the intractability of nature, but from insistence on retaining obsolete modes of expression that are scientifically impermissible, semantically not-well-formed, and philosophically idle.[2]

Einstein's own predictions seem to confirm the extensionalist in taking his "hard line." Is the contraction of a moving rod "real"? By the same token, is the slowing down of a moving clock "real"? The answer seems to be both yes and no. These phenomena are real from the point of view of one inertial observer who witnesses them, but not from the vantage point of another inertial observer who fails to notice them. Yet they cannot be called mere "appearances," for the following reasons: (1) since rest and uniform motion are in principle indistinguishable, the situation can be reversed, so that if A witnesses a dilation or a contraction on B's part, B can with as much justice say that it is occurring instead to A; (2) the word "appearance" suggests an underlying reality with which it can be compared, but, as (1) just made clear, no such reality is accessible to us in either special or, as it turns out, general relativity. The most we can say is that (using the Lorentz-Fitzgerald equation) we can measure what will happen to A as viewed by B, and vice versa, and that experiments will confirm our calculations in each case. And this is prcisely the extensionalist's point; if it seems like the least we can do, instead of the most, that is no doubt because we have failed to appreciate the capabilities, as opposed to the alleged limits, of science.

The curious thing is that Einstein's own views about science do not tally with this picture at all. Whereas the extensionalist in effect favors a plurality of descriptions of "reality," Einstein is always emphatic about the underlying unity of the physical world. In response to quantum physics, Einstein remarks that "the programmatic aim of all physics" is nothing but "the complete description of any (individual) real situation (as it supposedly exists irrespective of any act of observation or substantiation)."[3] Elsewhere he notes in very characteristic language that "the simpler our picture of the external world and the more facts it embraces, the stronger it reflects in our minds the harmony of the universe.[4]

Unlike the extensionalist, Einstein is something of a monist when it comes to describing the world. Both he and the extensionalist would agree on the presence of a plurality of entities (events, phenomena, states) within a given "field," but Einstein would say that there was one and only one set of forces governing that field; hence one and only one correct description of the field. Moreover, for Einstein the field is itself real, not just a "construct." Einstein and the extensionalist would again agree that it might be impossible for human beings ever to know the correct description of the field, but he would regard this as an epistemological (psychological, sociological) rather than as an ontological issue.

Viewed in this light, Einstein appears much less radical than his early reputation made him out to be. He is essentially a conserver of the Newtonian tradition of complete (deterministic) explanation of the universe. The extensionalist stands to Einstein in roughly the same way as Einstein stands to Newton: a more thorough-going version of the original theory necessitates a debunking of some of its main features. Just as Einstein retained the Galilean relativity principle and the constancy of the speed of light in order to do justice to a completely causal account of nature, so the extensionalist retains the notion of alternative and incomparable realities in order to do justice to the relativity of simultaneity and the shrinking of meter-sticks. Whereas the quantum physicist and Einstein are at odds, the extensionalist can lay claim to resolving the dispute in getting rid of classical conceptions that are held over, allegedly quite needlessly, in Einstein's theory of science. The fact that Einstein would be dissatisfied with this does not detract from his genius; it merely indicates, to the extensionalist, the failure of his philosophical thinking to catch up with his purely scientific innovations.

We find it hard to accept the conclusion just reached. But it is not enough to be guided by the intuitive belief that Einstein's insights <u>about</u> science must be at least as worthwhile or valid as his insights within science. What reasons can be adduced to support Einstein's classical emphasis on discovering the real? We would suggest several.

First, the thought experiments described above establish an irreducible difference between two sets of perceptions. But what is a perception or an experience? Perceptions are not encounters with "raw data" but are already subject to some sort of ordering arrangement or interpretation. This point is obviously Kantian in origin; it is expressed nowadays as the theory-ladenness of observation. But it is also point (2) of the original extensionalist doctrine, as set forth above. This means that the thought-experiments of Einstein cannot by themselves decide anything: the decision as to what is "seen" or what the seeing signifies, occurs at the level of theory. The extensionalist doctrine requires as much, yet it is curiously neglectful of same in drawing the conclusion that "reality" cannot be a univocal expression. If theoretical terms (and "reality" is surely one of these) are matters of arbitrary preference, then so is the decision to retain (or discard) the word "real" in its univocal sense, the sense which Einstein still attaches to it. Therefore, Einstein has as much right to hold an attitude as the extensionalist has to his, on extensionality's own grounds, no less.

Second, if this defense seems to render physics impotent to deal with nature, consider Einstein's remark on that subject:

> "Being" is always something which is mentally constructed by us, that is, something which we freely posit (in the logical sense). The justification of such constructs does not lie in their derivation from what is given by the senses. Such a type of derivation .... is nowhere to be had .... the justification of the constructs, which represent "reality" for us, lies alone in their quality of making intelligible what is sensorily given.[5]

This "quality of making intelligible what is sensorily given" is not just a euphemism for predictive power. It requires a set of laws which can serve, ideally, as "the presupposition of every kind of physical thinking," thereby "... making the totality of the contents of conciousness intelligible."[6] In other words, the thrust of his comments is once again classical or Newtonian in spirit: find the mathematically simplest set of covariant laws, laws adequate in principle for the description of all phenomena, now and in the indefinite future. Reality is mind-independent and stable; in this one respect Einstein is Parmenides to the extensionalist's Heraclitus. The independence of theory-construction from perceptual judgments is what allows science to function. Thus, puzzles arising at the level of data assimilation can, in principle, be resolved at the level of theory or hypothesis. None of this is intended to diminish the difficulty of actually doing so. What it does indicate is that only a philosophy still wedded to a kind of empiricism would conclude straightforwardly from the thought-experiments that questions about reality are either irresolvable or bogus. The freedom, as Einstein aptly puts it, to engage in theory-construction is what enables us to study nature; it is also what enables us to decide on the retention or rejection of a particular concept as part of the apparatus brought to bear on nature.

Third, of course, the extensionalist can always retort that "laws" of nature are nothing but chains of sentences occurring in a book or journal. But this is to ignore, if nothing else, their invariance in different coordinate systems, an invariance which is already demanded by adherence to the Galilean relativity principle, one of the two postulates of special relativity. (The principle of equivalence performs comparable functions in general relativity.) As Einstein (and every subsequent textbook) puts it, the laws of nature are independent of the choice of inertial system.[7] How can the extensionalist explain this? As a "mere" heuristic device? Not in light of its explanatory fruitfulness and (in the select cases in which it is as yet susceptible to test) its experimental verification. As a mere accident? That would revert, strangely enough, to the way in which Galileo himself saw the equivalence between gravitational and inertial mass.[8] He thought nothing more of it; it was this "omission" which prevented him from having some of Einstein's insights -- an omission of Newton's, as well. What was so remarkable about relativity was not the discovery of discrepancies between observers, but the claim that they did not matter (i.e., were not disastrous), because an area of agreement could also be found.[9] This area is expressed in equations like $E=mc^2$ and the Lorentz-Fitzgerald contraction formula. If a law is nothing but a chain of sentences, why do these work so well in such a variety of contexts? Even the probabilistic formulas of quantum mechanics are accurate to a very high degree, and can be said to account for nature, although Einstein rejected statistical explanation as an ultimate form of scientific thought.[10] Therefore, the extensionalist can find no comfort in the findings of Einstein's successors in physics, even though their disclosures about the interaction between observer and what is observed seem, at a superficial glance, to lend support to the extensionalist thesis that reality is a function of what is proximately measurable.

Fourth, it remains true that the extensionalist is committed to the use of some rational techniques of investigation. If different coordinate systems are irreducibly different from the perceptual standpoint, why bother to "measure" phenomena at all? Surely the acts of the scientist who "operates" under extensionalist inspiration must become ludicrous, if there is no assurance that methods employed lead, with sufficient diligence and persistence, to worthwhile results. This does not mean that such results are ever final and settled; that would violate the principle of fallibilism, which Einstein as much as any scientist upheld. Nor does it mean that the "assurance" in question is categorical, for that would violate Hume's injunction against the pretense to make empirical science demonstrative.[11] And, as we already noted, Einstein accepts the cognitive limitations of science, provided these are not misconstrued as ontological conditions. In any case there is an implicit realism in even the most thorough extensionalist approach. P. W. Bridgman, for example, while disparaging the way in which Einstein permits "a meaning ... [to be] assigned to the event in its own right apart from the frame of reference which yields the coordinates,"[12] nonetheless contends that "nature presents us with a unique frame of reference" and with "unique methods of description" for correlating and explaining data.[13] While he also remarks that "unique" methods of measurement "rob the contention that all frames of reference are equally significant of some of its intuitive appeal, "Bridgman's defense of Newtonian-style concepts robs his defense of operationism of all of its appeal.[14] While such an ad hominem circumstantial by no means justifies a rejection of extensionalilty, it may help to show that it is difficult, if not impossible, for the extensionalist to avoid committing himself to what are for him precritical notions about there being "a 'reality' back of all our multifarious experience."[15]

Fifth, this leads us to two related criticisms of the extensionalist viewpoint. One is, that it runs contrary to common sense. The other is, that it is repugnant to our deepest expectations. Neither argument by itself is either conclusive or weighty, yet each one depends on considerations frequently invoked on behalf of (or against) a point of view. The common sense argument is simple enough to state: if 'reality' simply means a set of operations, and the set is capable of yielding different results at different points in space-time, then the word "reality" has been blurred to the point of being meaningless. This, of course, is part of the extensionalist argument in favor of dropping the word, but it can be turned against the extensionalist himself. Common sense is, of course, a tricky thing. What seemed common sense to Newton was abhorrent to his contemporaries, viz., action at a distance. Common sense is continually being re-educated, and it may be that the extensionalist has lessons to teach us which simply have not yet been absorbed. On the other hand, we have seen that an extensionalist account of science seems to add to our difficulties in understanding the world, whereas we expect any explanation (both in science and philosophy) to reduce them.

Moreover, if extensionality is correct, hopes for a universally applicable set of laws must cease. What physicist acts on that hypothesis? What philosopher, despite the notorious difficulty of gaining agreement on such issues as free will or the existence of God? An extensionalist approach would counsel us to be modest in our expectations, but in doing so it seems to go overboard, in asking us, in effect, to abandon deeply cherished notions of uniformity and exceptionlessness. We long ago gave up trying to prove that such notions apply to nature, but that does not mean that we do not retain them as elements of method and as driving motives of research, as in Einstein's appeal to our sense of "harmony." Of course, the extensionalist has a right to see these as speculative and sentimental, but even this does not argue for their dispensability, only for their (admittedly widespread) subjectivity.

Finally, it can be said that the extensionalist, while he may condemn the retention of reality conceived as a regulative principle as merely speculative, has produced nothing but a series of speculations of his own, or more properly, a normative program which he wishes science to follow. This is perfectly legitimate, but it cannot legitimately exclude other normative programs from striving to install themselves as methodologically directive of science. This can be stated more forcefully: according to the extensionalist there is no reality per se, but only what is measurable. Is extensionality "measurable," is it a testable hypothesis? On its own terms, it is not and cannot be. Hence the choice in favor of extensionality is arbitrary and devoid of "evidence" in light of which it could be intelligently made. Of course, this is unfair to extensionalism in that it misrepresents it as not trying to account for what science does, but this misrepresentation is based on the extensionalist's own principles concerning decision making in science. If decision making in philosophy is to be conducted in another way, the extensionalist has failed to exhibit it. There is therefore a lack of what might be called dialectical depth to his position. As a program, it does not account for the facts of scientific inquiry, nor for the principles that govern scientific research. Additionally, it fails to be self-referential, to account for its own position in the constellation of possible (meta) theoretic attitudes, and thus it fails to provide grounds for its own adoption. Indeed, on its own terms it cannot even be comprehended.

The foregoing narrative provides some relief for the defender of Einstein, and it also shows that the appearance of rigor that the extensionality thesis conveys can be deceptive. So is the attempt to extrapolate extensionalist arguments from within the heart of relativity theory itself. However, this does not mean that Einstein's view of science is vindicated, either. Ideally, the format of this paper would have been as follows:

1. Reality is different (differently described) in various different coordinate systems. Events are nonsynchronous.[16]

2. This leads the extensionalist to say that different frameworks must all be incomparable, hence that physics or science in general must limit itself to describing frameworks one at a time.

3. But the extensionalist ignores the fact that the laws of nature are invariant, hence he runs afoul of the nonconventional and universal status of (mathematically based) predictions and equations.

4. Therefore, the extensionality thesis is wrong (and Einstein's account of the significance of his own work in science is fundamentally right).

Unfortunately, life is not as simple as this argument would make it out to be. While the extensionalist may have misinterpreted the status of laws, this does not mean that the notion of a mind-independent reality is unproblematic, as Kant and James, to name just two, have shown. Laws may be interpreted as fundamental relationships ingredient in the universe (Pierce), or else as tendencies to order implicit in the thinking process (Bergson). The issues involved in either conception are subtler and require more careful comparison than the scope of this paper allows.

But we can say this. Admittedly truth and falsity can sometimes be a matter of definition, but not always.[17] The valid parts of the extensionalist thesis reduce to the truism that all of our knowledge-claims are questionable, and that what we say is real determines, at any given moment, what is

real for us.   But we cannot <u>avoid</u> saying things about the
world, just because we happen to be part of it.   A spectator
of time and eternity like Plato's demiurge might be in a
position to know things which we are disbarred from knowing
on account of our active participation (transaction) with
things around us.   But this very act also puts us in touch
with the world, and what we say, including our most basic
physical posits, is a reflection of our attempts to get in
touch, to understand, to control, to survive.   Sometimes we
do not do it very well, but at no time are we uncommitted, in
the sense of having no defensible idea of what the world is
like.   A retreat from commitment is both false to fact and,
because it is a commitment, self-refuting.   These are mistakes which Einstein did not commit, because, among other
things, he <u>was</u> committed, to ideas which, whatever their
merit or ultimate tenability, are at least as defensible as
their extensionalist opposition.   Probably more so.

Postscript:

   Following our joint presentation at the centennial conference held at Hofstra University in November, 1979,
Professor Eugene Lashchyk of LaSalle College queried whether
we were "attacking a straw man" by making Einstein confront
an extensionalist.   Consider how ultra-sophisticated philosophy of science has become in the last few decades.   Mere
mention of the principal foes and adversaries (Kuhn,
Feyerabend, Popper, Lakatos, Agassi, Hanson, Toulmin,
Finocchiaro, Laudan, van Fraassen, Putnam, Stegmuller, Sneed,
Humphrey, Shapere, Watkins, Musgrave, Wartowsky, D. Miller,
H. Albert, Kordig) suffices to show that contemporary
dialectic has reached a high level of refinement.[18]   While on
imaginary debate between Einstein and Bridgman might have
historical value, can it claim to represent current thinking,
or is it merely out of date?   In response we contended that
subtle varieties of operationalism continue to plague both
philosophy of science and the natural sciences themselves.
But Lashchyk's direct challenge did compel us to find solid
(sociological) evidence in support of the claim that positivism is both a living option and a prevalent methodological
commitment.   Fortunately, the evidence is not hard to assem-
ble.   In philosophy, Glymour's "bootstrapping" version of the
deductive-nomological model of explanation[19] and Giere's
thoughtful defense[20] of the analytic tradition have restored
orthodoxy, while the dissenters have splintered, lost pres-
tige and gone into retreat.   Cartwright[21] goes so far as to
maintain a strict phenomenalism, denying that physics can
disclose underlying "patterns" in nature apart from the order
that mathematics imposes on (aspects of) experience.   While
her arguments are Aristotelian rather than Machian, the outcome is Kantian: what science describes is not the world but
our conventions for describing the world.   Davidson's notion
of a "conceptual scheme," Strawson's categorial analysis and
Goodman's austere semiotics[22] all reinforce the same conclusion.   These developments alike suggest that it is
Einstein's attempt to formulate a metaphysics of nature
rather than something descended from logical positivism which
seems quaint, old-fashioned and out of vogue.   Nor would many
scientists endorse (let alone carry out or try to ground)

Einstein's program for uniting quantum mechanics with general
relativity, by subsuming probability amplitudes under
(conjectured) causal relations.[23] Moreover, historian
Stanley Goldberg has shown just how the reception of relativity theory in the USA was thoroughly conditioned by opera-
tionalist expectations/preconceptions, which obscured and
distorted Einstein's original aims to the point where (as of
1980) they have never been recovered. Had Einstein's
American interpreters been aware of the metaphysical
character of what they (partially) absorbed, they would have
promptly rejected it as ungrounded speculation - thereby
changing the course of recent intellectual history as well as
postponing Einstein's canonization by the American press.[24]
To our knowledge, no one besides Goldberg has addressed these
issues, despite numerous volumes commemorating the centennial[25] and several significant reappraisals of Einstein's
thought[26] which have appeared in the seven years since we
wrote our essay. The subject is broader than either Lashchyk
or we imagined at that time: it concerns nothing less than
the formative assumptions which guide or govern Western
culture.[27] Whether we call this a paradigm, a Gestalt, a
conceptual scheme, a mind-set, a world-view, a semiotic code
or some other name matters little, just so long as we appreciate what is at stake. If our contribution goads cholars to
investigate Einstein's ontology more closely while spurring
further examination of the cultural matrix which shaped yet
never managed to accept it, we shall be gratified.

Notes

    1. A. Einstein and L. Infeld, The Evolution of Physics
(New York: Simon and Schuster, 1938), pp. 178 - 79.
    2. For an example of such impatience with "philo-
sophical" questions about reality, see N. D. Mermin's treatment of the Clock Paradox in chapter XVI of his Space and
Time in Special Relativity (New York: McGraw-Hill, 1968),
esp. p. 149.
    3. A. Einstein, "Remarks Concerning the Essays Brought
Together in this Co-Operative Volume" (hereafter 'Remarks'),
in Albert Einstein Einstein:Philospher-Scientist, 2 vols,
ed. P. A. Schilpp (Evanstan, Ill.: Library of Living
Philosophers, 1949 and subsequent editions), p. 667.
    4. Einstein and Infeld, The Evolution of Physics,
p. 213.
    5. A. Einstein, "Remarks," p. 669. There is a strongly
pragmatic flavor to this language, confirmed by what Einstein
says in other places, but even this is not what the extensionalist wants to hear and is quite distinguishable from the
latter's approach.
    6. Einstein, "Remarks," p. 673.
    7. Einstein, "Autobiographical Notes," in Albert
Einstein: Philosopher-Scientist, p. 57. Compare "Einstein's
gravitational field must ... be formulated without reference
to a unique coordinate system," from H. Reichenbach, The
Philosophy of Space and Time, tr. M. Reichenbach and J.
Freund (New York, Dover Publications 1958), p. 233.
    8. Einstein and Infeld, The Evolution of Physics, pp.
33 - 34, 215.
    9. Our thanks go to our colleague Dr. Stacy Langton

for calling this to our attention.

10. Einstein, "Remarks," pp. 672 - 73, and esp. p. 681.

11. For Hume's influence on Einstein, see the "Autobiographical Notes," pp. 13 and 53.

12. P. W. Bridgman, "Einstein's Theories and the Operational Point of View," in Albert Einstein: Philosopher-Scientist, p. 343.

13. Ibid. pp. 351 - 52.

14. Ibid. p. 351.

15. Ibid. p. 347, admittedly describing Einstein, not himself.

16. For an attack on sameness, see ibid., p. 346. For a defense of sameness, construed as "genidentity" of an event, see Reichenbach, The Philosophy of Space and Time, pp. 270 - 71.

17. On this issue, see H. Reichenbach, "The Philosophical Significance of the Theory of Relativity," in Albert Einstein: Philosopher-Scientist, p. 293. On Einstein's ontological commitments in general, see H. Margenau, "Einstein's Conception of Reality," in Albert Einstein: Philosopher-Scientist, Schilpp, p. 249.

18. For discussion and analysis see Frederick Suppe (ed), The Structure of Scientific Theories, 2nd ed (Urbana, Il: University of Illinois Press, 1977) and Harold I. Brown, Perception, Theory and Commitment: A New Philosophy of Science (Chicago: University of Chicago Press, 1979).

19. Clark Glymour, Theory and Evidence (Princeton, NJ: Princeton University Press, 1980).

20. Ronald Giere, Understanding Scientific Reasoning, 2nd ed. (New York: Holt, Rinehart and Winston, 1984). Also see Giere's important review article, "History and Philosophy of Science: Intimate Relationship or Marriage of Convenience?" British Journal for Philosophy of Science, Vol 24 (1973), 282 - 97. For a dissenting view (not directed at Giere), see L. Pearce Williams, "Should Philosophers be Allowed to Write History?" in ibid., Vol. 26 (1975), 241 - 53. For a provocative appraisal of what scientists do, and the discrepancy between real and "text-book" science, see Ian I. Mitroff, The Subjective Side of Science: A Philosophic Inquiry into the Psychology of the Apollo Moon Scientist (New York: American Elsevier, 1974).

21. Nancy Cartwright, How the Laws of Physics Lie (Oxford: Clarendon Press, 1983).

22. Donald Davidson, "On the Very Idea of a Conceptual Scheme" (1974), reprinted in Inquiries Into Truth and Interpretation (New York: Oxford University Press, 1984), Essay #13, pp. 183 - 98; P. F. Strawson, Individuals: An Essay in Descriptive Metaphysics (London: Methuen, 1959) and The Bounds of Sense: An Essay on Kant's Critique of Pure Reason (London: Methuen, 1966), Nelson Goodman, Languages of Art: An Approach to a Theory of Symbols (Indianapolis: Bobbs-Merrill, 1968) and Ways of Worldmaking (Indianapolis: Hackett Publ., 1978). The contributions of Ernst Cassirer and Susan Langer foreshadow these authors.

23. One physicist who does adhere to an Einsteinian world-view is Mendel Sachs. See General Relativity and Matter: A Spinor Field Theory from Fermis to Light Years (Dordrecht and Boston: D. Reidel, 1982) for his most ambitious attempt to follow through on Einstein's commitment to

classical determinism.

24. Stanley Goldberg, Understanding Relativity: Origin and Impact of a Scientific Revolution (Boston: Birkhaeuser, 1984), pp. 241 - 319 [pp. 293 - 305 trace Bridgman's four decades of struggle to comprehend Einstein]. Besides its masterful scholarship, Goldberg's work is arguable the clearest and most stimulating exposition of relativity theory for the non-specialist.

25. A. P. French (ed), Einstein: A Centenary Volume (Cambridge, MA: Harvard University Press, 1979); Gerald Holton and Yehuda Elkana (eds), Albert Einstein: Historical and Cultural Perspectives [The Centennial Symposium in Jerusalem 14 - 23 March 1979] (Princeton, NJ: Princeton University Press, 1982); Harry Woolf (ed), Some Strangeness in the Proportion: A Centennial Symposium to Celebrate the Achievements of Albert Einstein (Reading, MA: Addison-Wesley, 1980); Maurice Goldsmith, Alan Mackay and James Woudhuysen (eds), Einstein: The First Hundred Years (Oxford: Pergamon Press, 1980); Peter Barker and Cecil G. Shugart (eds), After Einstein: Proceedings of the Einstein Centennial Celebration at Memphis State University [14 - 16 March 1979] (Memphis, TN: Memphis State University Press, 1981); Mario Pantaleo and Francesco de Finis (eds), Relativity, Quanta, and Cosmology [in The Development of the Scientific Thought of Albert Einstein], Vols I-II (New York: Johnson Reprint Corp., 1979); Peter C. Aichelburg and Roman U. Sexl (eds.), Albert Einstein: His Influence on Physics, Philosophy and Politics (Braunschweig and Vienna: Friedrich Vieweg & Son, 1979).

26. Abraham Pais, Subtle is the Lord: The Science and the Life of Albert Einstein (New York: Oxford University Press, 1982); Lewis S. Feuer, Einstein and the Generations of Science, 2nd ed (New Brunswick, NJ: Transaction Books, 1982); Loud Swenson, Jr., Genesis of Relativity: Einstein in Context (New York: Burt Franklin & Co., 1979); Albert I. Miller, Albert Einstein's Special Theory of Relativity: Emergence and Early Interpretation 1905 - 1911 (Reading, MA: Addison-Wesley, 1981). For broader historical assessments of Einstein's accomplishments, see esp. I. Bernard Cohen, Revolution In Science (Cambridge, MA: Harvard University Press, 1985), and Gerald Holton, Thematic Origins of Scientific Thought (Cambridge, MA: Harvard University Press, 1973).

27. See Carolyn Merchant, The Death of Nature: Women, Ecology and the Scientific Revolution (New York: Harper & Row, 1980); Brian Easlea, Witch Hunting, Magic and the New Philosophy: An Introduction to Debates of the Scientific Revolution (Atlantic Highlands, NJ: Humanities Press, 1981); Allan G. Debus, Man and Nature in the Renaissance (Cambridge: Cambridge University Press, 1978). E. J. Dijksterhuis' The Mechanization of World Picture, trans. C. Dikshoorn (Oxford: Clarendon Press, 1961) and Richard S. Westfall's The Construction of Modern Science: Mechanisms and Mechanics (Cambridge University Press, 1978) are equally authoritative accounts of the science involved in the scientific revolution.

Part VI

Literature

13.

"Springtime of the Mind": Poetic Responses to Einstein and Relativity

CAROL DONLEY

Stephen Leacock recalls that in 1923 at the height of popular interest in Einstein, he asked Ernest Rutherford

> what he thought of Einstein's relativity. "Oh, that stuff!" Rutherford replied. "We never bother with that in our work!" . . . When the German physicist Wien told Rutherford that no Anglo-Saxon could understand relativity, Rutherford replied, "No, they have too much sense." But it was Einstein who made the real trouble. He announced in 1905 that there was no such thing as absolute rest. After that there never was."[1]

The poetic responses to relativity appear in two ways: first, writers used Einstein and concepts of relativity as content or subject matter of their poetry; second, writers invented formal or structural analogues to relativistic space and time. This chapter examines both poetic responses to relativity and suggests that the physicists and poets are not always so diametrically opposed as conventional criticism assumes.

Two significant poets to select Einstein as protagonist in their poems were William Carlos Williams and Archibald MacLeish, both of whom read Whitehead's Science and the Modern World and studied the new physics with care. Williams' first poetic response came in his three page poem, "St. Francis Einstein of the Daffodils," which he published in the magazine Contact in 1921 shortly after Einstein's first visit to the United States. The poem celebrates Einstein as a saint and liberator bringing "springtime of the mind," "bringing April in his head."

> April Einstein
> through the blossomy waters
> rebellious, laughing
> under liberty's dead arm
> has come among the daffodils
> shouting
> that flowers and men
> were created
> relatively equal.[2]

All the imagery of the poem promises a new birth as "old-fashioned knowledge is/ dead under the blossoming peach-trees."   In fact, for Williams and many others, Einstein brought fresh hope.   Not only did he show how to go beyond the old classical conventions, but he opened up a whole new world of possibilities for artists as well as for physicists.

One member of the Paris avant garde during the mid-1920s was Archibald MacLeish, who published his long poem "Einstein" in 1926.   Hyatt Howe Waggoner considers this poem to be "not only the finest poetic tribute to the scientist, [but] also an informed and interesting comment on the philo-sophical significance of Einstein's achievement."[3]

The poem "Einstein"[4] traces the steps of the scientist's inquiring mind as it moves from contemplating his own finite limits and conflicting concepts of reality to the mind's achievement of the solution.   After rejecting

> A world in reason which is in himself
> And his own dimensions

Einstein tries to find ways to articulate his thoughts. Models cannot express his thinking; music "vaguely ravels into sound." Words cannot convey his ideas.   "There is no clear speech that can resolve/ Their texture to clear thought ... Now there are no words/ Nor names to name them." Moving into further abstractions, Einstein

> lies upon his bed
> exerting on Arcturus and the moon
> forces proportional inversely to
> The squares of their remoteness and conceives
> The Universe.
>   Atomic
> Oceans in atoms and weigh out the air
> In multiples of one and subdivided
> Light to its numbers.
>   If they will not speak
> Let them be silent in their particles.
> Let them be dead and he will lie among
> Their dust and cipher them — undo the signs
> Of their unreal identities and free
> The pure and single factor of all sums —
> Solve them to unity.[4]

As Waggoner remarks, "The chief idea to emerge from the poem is that the new Einsteinian science has at once empha-sized the centrality and increased the loneliness of the knowing of the mind."[6]

The sense of an isolated intelligence in a vast, indif-ferent cosmos also appears in a Robert Frost poem, only Frost's antiscience sarcasm is poles apart from MacLeish's informed respect for the scientist.   In Frost's poem, the loneliness is associated with an infinite Newtonian universe, but the cozy, curved space seems to comment unfavorably on the mediocrity of the contemplating mind.

Any Size We Please

No one was looking at his lonely case:
So, like a half-mad outpost sentinel,
Indulging an absurd dramatic spell,
Albeit not without some shame of face,
He stretched his arms out to the dark of space
And held them absolutely parallel
In infinite appeal. Then saying "Hell,"
He drew them in for warmth of self-embrace.
He thought if he could have his space all curved,
Wrapped in around itself and self-befriended,
His science needn't get him so unnerved.
He had been too all out, too much extended.
He slapped his breast to verify his purse
And hugged himself for all his universe.[6]

Frost's attitude towards science was ambivalent. As Lawrance Thompson comments, science simultaneously fascinated and repulsed Frost. "He was always deeply absorbed by details concerning any new scientific discovery. At the same time, he could be offended by any cocksure scientific manner which [he felt mocked] poetic and religious concerns for true mysteries."[7] When, for example, Frost heard lectures by Niels Bohr and had dinner with him in Amherst in 1923, the poet was intrigued and full of admiration; but when, a few years later, he had dinner with Robert Millikan and J.B.S. Haldane at the California Institute of Technology, the poet was rude and disrespectful, claiming "that their thinking was mere metaphor-making — and poor metaphor-making."[8] The same ambivalence often appears in his poetry, which shows both an awareness and a rejection of scientific concepts as, for example, in the poem "Skeptic." There Frost states, "I put no faith in the seeming facts of light" and "I don't believe what makes you [a star] red in the face/ Is after explosion going away so fast."[9]

Although very different from Frost in most respects, Ezra Pound resembled him in his attacks on science. Pound, a moral absolutist, did not reject Einstein the man or his physical theories, but rather the "socialization" of relativity — "the conception it has given rise to in the minds of laymen and many philosophers alike that 'everything is relative' and there are no absolute standards by which men ought to guide their moral responsibility."[10] Pound felt that Einstein, too, rejected such unwarranted extension of a purely physical theory into ethics and social behavior. As he said in his Guide to Kulchur, "Al Einstein scandalized the professing philosophists by saying, with truth, that his theories had no philosophic bearing."[11]

E. E. Cummings, too, in spite of his avant garde experiments with form, maintained a romantic hostility to science and scientists. Sometimes his criticism seemed gentle, as in his remark, "I have been found guilty of the misdemeanor known as ... making light of Einstein."[12] More often, however, he saw science as exploiting, depersonalizing, levelling, and destructive. "So far as I am concerned [he wrote] mystery is the root and blossom of eternal verities while, from a scientific standpoint, eternal verities are nonsense & mystery is something to be abolished at any cost."[13]

Cummings' poem "Space being (don't forget to remember) Curved" exemplifies his attitude:

    Space being(don't forget to remember)Curved
    (and that reminds me who said o yes Frost
    Something there is which isn't fond of walls)

    an electromagnetic(now I've lost
    the)Einstein expanded Newton's law preserved
    conTinuum(but we read that beFore)

    of Course life being just a Reflex you
    know since Everything is Relative or

    to sum it ALL Up god being Dead(not to

    mention inTerred)
                    LONG LIVE that Upwardlooking
    Serene Illustrious and Beatific
    Lord of Creation,MAN:
                    at a least crooking
    of Whose compassionate digit,earth's most terrific

    quadruped swoons into billardBalls![14]

This fractured and rearranged sonnet echoes Pound's complaint about the careless extension of relativity into standards for behavior. Note that "god" is in small case while "Lord of Creation, MAN" receives typographical emphasis. Cummings comments that achievements in abstract physical concepts have, in effect, reduced man's humanity. While Michaelangelo's God on the Sistine ceiling creates man with finger touch, man's "compassionate digit" pulls the trigger and kills for ivory. Man is not a safe or wise god.

By no means did all contemporary poets share Frost's, Cummings', and Pound's feelings. Many, in fact, found concepts in the new physics which seemed to offer metaphysical justifications for their experiments in literature. The poet, Louis Zukofsky, who translated Anton Reiser's biography of Einstein, said that the aims of poetry "and those of science are not opposed or mutually exclusive; and that only the more complicated, if not finer, tolerance of number, measure and weight that define poetry make it seem imprecise as compared to science."[15] In a similar way, William Carlos Williams asked, "How can we accept Einstein's theory of relativity ... without incorporating its essential fact — the relativity of measurements — into our own category of activity? Do we think we stand outside the universe? ... Relativity applies to everything."[16] Since a poetic line is a measure or group of measures, many poets believe it should be relativized — the measure of the line should be relative to the line's "reference frame." One contemporary aesthetic emerging from this is Charles Olson's statement (which he borrowed from Robert Creeley and passed on to William Carlos Williams) that form is never more than an extension of content. In other words, the poet no longer fits his content into sonnets or quatrains, but instead allows the form and content to emerge together as an organic whole.

These innovations in form shocked many who expected traditional structures in poetry and who would agree with Frost that writers experimenting with so-called "free verse" were playing tennis with the net down. Literary critics often describe the contrast between conventional form and more experimental modern structures as the difference between closed and open form. Poems written in closed form refer to governing principles external to the work, principles of meter, rhyme, stanzaic pattern such as the fourteen line iambic pentameter of the English sonnet with its four cluster rhyme scheme, as in Frost's poem, "Any Size We Please."

Open form, on the other hand, refers to poems whose structures are relative to the content of the poem. Both Williams' "St. Francis Einstein of the Daffodils" and MacLeish's "Einstein" exemplify open form.

In conclusion, it is interesting to note that literary critics are not alone in assuming a gulf between physics and poetry. Several texts popularizing physics for the layman also make that assumption. In fact, Baker's Modern Physics and Anti-Physics and March's Physics for Poets both imply that if poets can understand these explanations, anybody can. Clearly, however, many modern American poets have paid attention to the developments in twentieth-century physics; and several poets, Williams, MacLeish and Zukofsky, in particular, honored and respected Einstein as a great scientist whose ideas opened up a new world of aesthetic possibilities for the artists.

Notes

1. Stephen Leacock, "Common Sense and the Universe," in The World of Mathematics, vol. 4 ed. James R. Newman, (New York: Simon and Schuster, 1956), p. 2465.

2. William Carlos Williams, "St. Francis Einstein of the Daffodils," Contact 4 (1921): 2 - 4.

3. Hyatt Howe Waggoner, The Heel of Elohim: Science and Values in Modern American Poetry (Norman: University of Oklahoma Press, 1950), p. 144.

4. Archibald MacLeish, "Einstein," in Streets in the Moon (Boston: Houghton Mifflin, 1926). Also in Poems: 1924 - 1933 (Boston: Houghton Mifflin, 1933), 67 - 75.

5. Waggoner, The Heel of Elohim, p. 144.

6. Robert Frost, The Poetry of Robert Frost (New York: Holt, Rinehart, and Winston, 1969), p. 396.

7. Lawrance Thompson, Robert Frost: The Years of Triumph, 1915 - 1938 (New York: Holt, Rinehart and Winston, 1970), p. 288.

8. Thompson, Robert Frost, p. 658.

9. Frost, The Poetry of Robert Frost, p. 389.

10. David D. Pearlman, The Barb of Time: On the Unity of Ezra Pound's Cantos (New York: Oxford Univ. Press, 1969), p. 204.

11. Ezra Pound, Guide to Kulchur (London: Faber & Faber, 1938), p. 34; rpt. 1952, Norfolk.

12. E. E. Cummings, letter to Kenneth Burke, in Selected Letters of E. E. Cummings (New York: Harcourt, Brace, and World, 1969), p. 248.

13. Letter to Eva Hesse, in Selected Letters, p. 265.

14. E. E. Cummings, "Space being (don't forget to remember) Curved," The Norton Anthology of Modern Poetry,

ed. Richard Ellman and Robert O'Clair (New York: W. W. Norton Co., 1973), p. 535. Originally published in W: Seventy New Poems, 1931.

15. Prepositions: The Collected Critical Essays of Louis Zukofsky (New York: Horizon Press, 1968), p. 15.

16. William Carlos Williams, "The Poem as a Field of Action," in Selected Essays of William Carlos Williams (New York: Random House, 1954), p. 283.

14.

The Circuitous Path:
Albert Einstein and the
Epistemology of Fiction

ROBERT HAUPTMAN and IRVING HAUPTMAN

> When I examine myself and my methods of thought I
> come to the conclusion that the gift of fantasy
> has meant more to me than my talent for absorbing
> positive knowledge.
>
> Albert Einstein

The zeitgeist and the general inferences drawn from Einstein's work come to bear most seminally on a group of philosophically oriented novelists conveniently termed absurdists, authors who believe that man is, in Heidegger's phrase, "thrown" into a world devoid of absolutes, order, and meaningfulness. The world that these novelists depict is one in which external meaning is elusive, metaphysical underpinnings are questioned, and knowledge is ephemeral. If this is the antipode of the harmonious world that Einstein demanded spiritually, it is nonetheless a valid hypostatization that follows logically from his physical theorizing. It is small consolation that the man who helped to destroy the harmony of the universe, failed to accept the consequences of his own work and spent the rest of his life attempting to prove that the universe is indeed harmonious and comprehensible. Thus to ascribe man's anguished cry entirely to a philosophical etiology, while conveniently ignoring concomitant advances in physics is rather naive. Franz Kafka, Jean-Paul Sartre, Albert Camus, and Samuel Beckett were certainly influenced by nineteenth-century philosophy but their fiction consistently reflects the world that Einstein and Heisenberg depict. A. A. Robb commenting on only one aspect, Einstein's theory of simultaneity, states that "This seemed to destroy all sense of the reality of the external world and to leave the physical universe no better than a dream, or rather a nightmare."[1] Exactly. For Einstein's heirs nothing remains stable and the inevitable result is absurdity.

The first of the absurdists is Kafka, who, in 1910, became aware of relativity through direct contact with Einstein at the salon of Berta Fanta in Prague: "... he got familiar, shortly before writing his main works, with the most important matters of inquiry of the new age (for example, quantum theory and relativity theory."[2] Moreover, both Sartre and Camus acknowledge their debt to Kafka.[3] These are admittedly tenuous connections, but they certainly demand recognition. Furthermore, this chapter does not contend that these writers were directly influenced by Einstein's work, that there is an explicit line traceable from his scientific papers to their fictional worlds. It is rather claimed that during a period that witnessed the transformation of the most basic physical concepts, it is the general zeitgeist that exercised the primary influence on the absurdists. One of the few critics to note this is S. Beynon John, who, in discussing the intellectual climate of Camus's early years, observes that,

> New scientific theories, too, seemed to challenge still further men's assumptions about the nature of experience. Among these we must count the delayed implications of Freud, and discoveries about the nature of the physical universe, especially those of Einstein. Two general conclusions were often drawn from the play of these factors. Firstly, they appeared to break up traditional values and beliefs about the nature of man and his place in the universe. Next, in the degree to which they menaced individuality or made it the prey of unconscious impulses (as with Freud), these forces seemed to impair the density of individual existence and to provoke the idea that man was adrift in an absurd universe.[4]

It is this general Einsteinian influence that is so important for fiction.[5]
Because Einstein's scientific methodology is based on a surprisingly well-developed epistemology and more significantly because in an indeterminate and meaningless universe epistemological questions assume unprecedented importance, it is legitimate to examine the attempts of fictional characters to acquire knowledge. But it is not the purpose of this chapter to discuss Einstein's epistemology nor to compare his perspective with ways of knowing in fictive worlds. These are autonomous realms.[6]

> Time present and time past
> Are both perhaps present in time future,
> And time future contained in time past.[7]

These opening lines of Eliot's "Burnt Norton" mirror the confusion of time in modern fiction: from Proust, Joyce, and Mann through Robbe-Grillet and Cortazar, time is manipulated into asymmetrical, simultaneous, and reversible patterns. Events are not disclosed in sequential narrative, one incident succeeding another, but rather "everything is everywhere at all times." Rudolf Arnheim perceptively remarks that

> The shattering of the narrative sequence challenges the reader or viewer to reconstruct the objective order of the events. In trying to do so, he tends to assign the scattered pieces to their place in a structurally separate system offered by Time and Space. However, if the reader or viewer would limit his effort to this reconstruction of objective reality, he would

miss the entire other half of the work's structure. Although discontinuous and therefore dis-
orderly with regard to objective reality, the
presentation must also be understood and accepted
as a valid sequence of its own, a flow of disparate fragments, complexly and absurdly related
to one another.[8]

Strangely enough, the novels of Kafka, Sartre, Camus, and
Beckett frequently do unfold in a sequential fashion; that
is, a distinct temporal progression can occur. On the other
hand, time, for these writers, tends to be inconsequential,
amorphous, alienating. As Kafka notes in his diary, "The
clocks are not in unison, the inner one runs crazily on ...
the outer one limps along at its usual speed. What else can
happen but that the two worlds split apart ... ?"[9] This dis-
parity reinforces the chaos of a world in which the most
significant things are unknowable and even the insignificant
often defy comprehension. Attempts to discover are met with
silence. Indeed, Camus insists that "The absurd is born of
this confrontation between the human need and the unreason-
able silence of the world."[10] If one cannot know, then one
cannot predict and if one cannot predict, it is difficult to
know: this is a vicious circle and thus it is no surprise
that the fictional worlds created here are haphazard, dis-
harmonious, and unpredictable.

Both Kafka's Trial and Camus's Stranger are predicated
on unpredictable events and unfold through the unknowable.
This is, in fact, Kafka's normal perception of the world and
there are potent auguries of the weltanschauung of The Trial
and The Castle in Kafka's shorter fiction. In "The Judgment"
George commits suicide because his father sentences him to
death, but the sentence is incommensurate with his crime,
which is not really knowable, particularly since it may be
different for father, son, and reader. "In the Penal Colony"
is a detailed account of a grotesque device used for tortur-
ing prisoners, but the offender who is about to be tormented
does not know his sentence; in fact, he does not even know
that he has been sentenced. Joseph K., in The Trial, is
arrested, harassed, and ultimately executed (murdered by
official thugs) without ever discovering his crime; as K.
laments, "it is an essential part of the justice dispensed
here that you should be condemned not only in innocence but
also in ignorance."[11] The Trial is K's valiant attempt to
understand his position in a bizarre world, to learn what he
has done, and, more important, to discover how to deal
effectively with the judicial system. To this end he will
try anything, but everything fails; he learns nothing of
utility, and because of this he dies: knowledge is life;
ignorance is death.

Kafka retells the story in The Castle. The orientation
is different, but the goal is the same: to learn and to
understand, both of which are allegorized by K's attempt to
reach the castle. Of course, he never does; too many obsta-
cles are put in his path. He fails to recognize his assistants, who have supposedly followed him to the new town. This
is no surprise since they appear to be local men, not his
original assistants at all. K. accepts this and acts as if
they were the originals despite the evidence: "'But if you

are my old assistants you must know something about it
[surveying],' said K.   They made no reply.   'Well, come
in....'".[12]   The only reliable source of information is the
meaningless "humming and singing" of the telephone; "Everything else is deceptive."[13]   Such a world can hold few surprises for an acclimatized reader and therefore letters that
praise for undone work and confusions concerning identity are
taken for granted, even though Kafka expends multitudes of
words constructing labyrinthine arguments about them.   Franz
Kuna epitomizes Kafka's unique position:

> The Trial and The Castle are monuments to Kafka's
> dedication to his self-imposed task of dismant-
> ling the key assumptions underlying the idea of a
> harmonious order.   What Einstein very much at the
> same time, did in physics Kafka did in the field
> of ethics and aesthetics.[14]

Camus constructs The Stranger along similar lines:
Meursault knows why he is condemned, but he does not know why
he kills the Arab.   And the judges, jury, and prosecutor, who
think they know why Meursault is guilty, confuse the crime
with his social ineptitudes.   The Stranger, in a sense, is
Camus's paean to uncertainty: "Mother died today.   Or, may-
be, yesterday; I can't be sure."[15]   From these opening words
to the final execration, The Stranger insists that it is virtually impossible to know.   Even the commonplace is quali-
fied: phrases like "I suppose," "So far as I knew," "I
couldn't say," "I wasn't sure," "I can't remember," "It seemed," "I didn't know," and "I still don't know" appear in unabashed profusion.   And even when Meursault appears to know,
he actually does not, for example, when he acts as a witness
for Raymond, he is merely repeating something that Raymond
told him; but he does not check Raymond's story: it may be
true, it probably is, but Meursault has no way of knowing.
When he kills the Arab, he is dealing with a "blurred dark
form" and his eyes are filled with brine and tears so that he
cannot see.   He shoots and then shoots four more times.   He
does not know why he does so.   At his defense, the best he
can do is to mention the sun.   Meursault concludes by opening
his heart "to the benign indifference of the universe."[16]
For a man who neither knows nor cares to know, this is perhaps the only solution.
     The extremes of epistemological questioning are evident
in Sartre's Nausea and Beckett's The Unnamable.   Sartre is
more concerned with the metaphysical perspective.   The nausea
that Roquentin feels whenever he perceives too lucidly is
Sartre's metaphor for the congruency of knowledge and absurdity: the implied meaninglessness of The Stranger becomes
explicit in Roquentin's articulated responses to his environment.   Without self-knowledge and unsure of his purpose,
Roquentin abandons his historical study of Rollebon, about
whom little is known with certainty.   Nausea is of particular
interest in this context, because it is here that Sartre
presents one of the few characters in absurdist fiction who
actually has apodictic knowledge of the universe.   But the
self-taught man, who has acquired all of his knowledge by
reading the books on the local library's shelves in alphabet-
ical order (he is only half-way through), is a parody, a man

for whom knowledge is of little significance:

> He has digested anti-intellectualism, manicheism,
> mysticism, pessimism, anarchy and egotism: they
> are nothing more than stages, unfinished thoughts
> which find their justification only in him.[17]

Although Roquentin sees a solution to his dilemma in fictional creation (he will become a story teller) the final
effect of Nausea is mitigated by Roquentin's belief that,
"Every existing thing is born without reason, prolongs itself
out of weakness and dies by chance."[18]

In Molloy and Malone Dies Beckett depicts a quest whose
epistemological significance is invariably overshadowed by
metaphysical queries. But by The Unnamable (the final novel
in the trilogy), the inexorable movement has been reversed
and the metaphysical aspects give way to a 123 page litany of
man's inability to know himself, his past, his desires, his
physical surroundings, or his world. There are few pages
upon which the narrator fails to mention his lack of knowledge and whatever he does claim to know is immediately con-
tradicted. Metaphysical conjecturing breaks down in this
epistemological quagmire. At the height of his despair
Roquentin cries,

> I am. I am, I exist, I think, therefore I am; I
> am because I think, why do I think? I don't want
> to think any more, I am because I think that I
> don't want to be, I think that I ... because...
> Ugh! I flee.[19]

A stronger congruence of fictional worlds would be difficult
to imagine. This passage provides a gloss to the Unnamable's
plight. Roquentin reaches this stage and then moves on to
recovery. The Unnamable begins and ends in similar mental
gyrations. Since physically he consists (apparently) of a
torso embedded in a container, there is little he can do
other than think. But his goal is to "go silent," although
he begins by affirming that "I shall never be silent.
Never."[20] To which he adds, some pages later, "So it is I
who speak, all alone, since I can't do otherwise. No, I am
speechless."[21] He knows nothing and the reader knows even
less, because it is impossible to distinguish between the
truth and his mistakes, invented memories, tergiversations,
and lies. No cartographer could map the narrator's progress;
there is none. Early in the monologue he depicts himself at
rest or in motion, the distinction is unimportant, while
toward the end he laments that he is still unsure of what he
is, where he is, whether vocal or silent, indeed whether he
even exists. George Wellwarth sums up Beckett's position:
"all knowledge is an illusion and all things are pointless
— insofar as the human mind is concerned."[22] This is the
obvious conclusion and it is therefore easy to forget after
all those devastating words, that The Unnamable ends on a
positive note: "it will be I, it will be the silence, where
I am, I don't know, I'll never know, in the silence you don't
know, you must go on, I can't go on, I'll go on."[22]

The four novelists discussed here depict worlds in which
meaninglessness is the dominant value and the validity and

significance of knowledge is highly questionable. It is the
contention of this essay that this perspective can, in part,
be ascribed to developments in modern science, particularly
to Einstein's theorizing. That these developments logically
lead to the cul-de-sac outlined above has at least been noted
by the humanists. Few scientists, however, have ventured
such opinions. An exception is P. W. Bridgman, the Nobel -
Prize winning physicist, who in 1950 stated that,

> We are now approaching a bound beyond which we
> are forever estopped from pushing our inquiries,
> not by the construction of the world, but by the
> construction of ourselves. The world fades out
> and eludes us because it becomes meaningless. We
> cannot even express this in the way we would
> like. We cannot say that there exists a world
> beyond any knowledge possible to us because of
> the nature of knowledge. The very concept of
> existence becomes meaningless. It is literally
> true that the only way of reacting to this is to
> shut up. We are confronted with something truly
> ineffable. We have reached the limit of the
> vision of the great pioneeers of science, the
> vision, namely[,] that we live in a sympathetic
> world, in that it is comprehensible by our
> minds.[24]

If Einstein's influence on the absurdists is, at times,
oblique, there is another area of fiction where it is far
more salient. Many contemporary novelists have been capti-
vated by various aspects of modern science and they make use
of scientific method and metaphor often in strange and even
grotesque ways.[25] Perhaps the favorite metaphor to be usurp-
ed from its rightful place in the physicist's arsenal is
entropy. As Tony Tanner points out, the concept is virtually
ubiquitous and Norman Mailer, Saul Bellow, and John Updike,
inter alios, actually use the term in their fictions.[26]
Because entropy entails not merely the running down of the
universe, but disorder and chaos as well, it is necessary to
mention a specific manifestation. Thomas Pynchon's Crying of
Lot 49 revolves around the entropy of communication. Oedipa
Maas attempts to solve the mystery of an alternate mail
system, but she never learns whether the source of her knowledge is reality, hallucination, fantasy, or an extravagant
perpetration; the novel concludes in ambiguity and the ultimate effect, though not as powerful, is similar to that
achieved by Kafka and his heirs: one only discovers that one
cannot know with certainty.
  The four novels of Lawrence Durrell's Alexandria Quartet
provide an excellent example of the novelist using an
Einsteinian metaphor as a structuring principle. As Durrell
puts it in his note to Balthazar,

> Modern literature offers us no Unities, so I have
> turned to science and am trying to complete a
> four-decker novel whose form is based on the
> relativity proposition.
> Three sides of space and one of time constitute
> the soup-mix recipe of a continuum. The four

novels follow this pattern.
The three first parts, however, are to be
deployed spatially (hence the use of 'sibling'
not 'sequel') and are not linked in a serial
form. They interlap, interweave, in a purely
spatial relation. Time is stayed. The fourth
part alone will represent time and be a true
sequel.[27]

The result of this can be seen in the final novel, Clea. As
in Faulkner's Sound and the Fury, events are surprised from
different angles, which shows once again that knowledge is
subjective and ephemeral. The philosophical position is
articulated at various points in terms of "the mutability of
all truth" or at its most extreme "poetic or transcendental
knowledge somehow cancels out purely relative knowledge"[28]
More important, however, are the many examples of inexacti-
tude: Capodistria's death, which turns out to be a hoax;
Darley's vision of Justice, which is "an illusionist's creation;" the incredible transformation of Scobie into the
Saint, El Yacoub; or the misconceptions concerning Purse-
warden, who is perceived from the perspectives of Darley, the
diary, his sister, his wife, and Keats. The only knowledge
possible here is rather peculiar: "Sexual love is knowledge,
both in etymology and in cold fact ... When a culture goes
bad in its sex all knowledge is impeded."[29] While Einstein
believes that there are no limits to knowledge, Durrell,
using an Einsteinian structure, concludes that, with the
exception of sexual knowledge, epistemological problems are
unresolvable; as he observes in another context, "Under the
terms of the new idea a precise knowledge of the outer world
becomes an impossibility."[30]
Stanislaw Lem's protagonists, like Kafka's, are usually
searching for the solution to some enigma. The Investigation
provides a splendid example of characters who attempt to discover order and harmony in apparent chaos. The disappear-
ances of some corpses result in a number of tangential
explanations, the most preposterous of which are couched in
purely statistical terms. The result of one analysis, "the
product of the distance and the time between consecutive
incidents, multiplied by the temperature differential" is a
constant and leads nowhere.[31] A second statistical farrago
insists that the answer to the mystery lies in the fact that
the corpses disappeared in the area of England that has the
lowest cancer death rate. The novel concludes with yet
another hypothesis concerning truck drivers, fog, and fanta-
sies. When confronted with this explanation, Gregory asks,
"is all this true?" to which the Chief Inspector replies, "No
but it might be. Or, strictly speaking, it can become the
truth."[32] These men are groping for knowledge in a world in
which bodies cannot be identified with certainty; time
factors are ambiguous; rumors are rampant; actions are incom-
prehensible; and a guilty party may not exist. Nonetheless
they continue to insist on specious explanations for an
unsolvable case. As Gregory declares in a moment of despair,
"we human beings are the resultant of Brownian motion ... Our
knowledge is underlined by statistics — nothing exists except
blind chance, the eternal arrangement of fortuitous
patterns."[33] It is perhaps superfluous to add that The

Investigation concludes in ambiguity; there is no solution.
Concerning the reliance of statistical methods of prediction, Philipp Frank remarks, "If science could not advance
beyond this stage, 'God would,' as Einstein said, 'play dice
indeed.'"[34]   Robert Coover has predicated his fascinating
novel, _The Universal Baseball Association, Inc. of J. Henry
Waugh, Prop._, on this supposition.   J. Henry Waugh (whose
initials plus the final h form the letters of the Biblical
name for God, JHWH) controls a baseball game of his own creation, by rolling dice.   Waugh's involvement in this board
game is so intense that he fails to distinguish between it
and reality; thus he alienates friends, loses his job, and
imperils his own being.   The game progresses according to the
throw of the dice and until Damon Rutherford, Waugh's
favorite player, is killed by a pitch, Waugh does not interfere.   There is nothing that he can do to bring Rutherford
back, but he does meddle in future games; he juggles sched-
ules; he controls the Knickerbockers' losses; and he tampers
with two rolls of the dice, which results in the death of
Damon's killer.   In Arlen Hansen's apt words, "God _does_ play
dice with the universe, but the dice are loaded."[35]   The
novel concludes in a ritual reenactment of Damon's death.
This mythification is Coover's subtle indication that one
knows only in confusion: the player who becomes Rutherford
in the reenactment is not really the earlier hero, but in the
eyes of the crowd he _is_, and as such he must be sacrificed,
literally.   The ambiguity of the final pages allow for a
tantalizing peroration: the reader learns nothing further of
this ball player's fate, nor of Waugh's for that matter.
The preceding discussion attempts to show that
Einstein's influence on novelists is both substantial and
diverse.   He is indirectly responsible for the world that the
absurdists depict, the philosophical implications of which
have had a decisive impact on the contemporary mood.   Secondly, he directly influenced writers like Durrell and Coover
whose fiction depends both structurally and contextually on
Einsteinian metaphor.   Finally, there are novelists like
Vladimir Nabokov and Aldous Huxley, who merely mention
Einstein or his theories in passing.   Further, this essay
contends that many of these novels lead to the same conclusion: in an Einsteinian universe knowledge is ephemeral,
elusive, and at times unobtainable.

Notes

1.   R. W. Clark, _Einstein: The Life and Times_ (New York:
World Publishing Co., 1971), p. 243, quoting A. A. Robb.
2.   Bert Nagel, _Franz Kafka: Aspekte zur Interpretation
und Wertung_ (Berlin: Erich Schmidt Verlag, 1974), p. 319,
quoting Klaus Wagenbach.
3.   See Maja J. Goth, "Existentialism and Franz Kafka:
Jean-Paul Sartre, Albert Camus and their Relationship to
Kafka," in _Proceedings of the Comparative Literature
Symposium; Franz Kafka: His Place in World Literature_, ed.
Wolodymyr T. Zyla (Lubbock: Texas Tech University, 1971), IV,
51-69.   And Albert Camus, "The Myth of Sisyphus," in his _The
Myth of Sisyphus and Other Essays_, trans. Justin O. O'Brien
(New York: Vintage Books, 1955), especially pp. 92 -102, an
appendix entitled "Hope and the Absurd in the Work of Franz

Kakfa."

4. S. Beynon John, "Albert Camus," in On Contemporary Literature, ed. Richard Kostelanetz (New York: Avon, 1964), p. 307.

5. Alan Warren Friedman, Lawrence Durrell and the Alexandria Quartet (Norman: University of Oklahoma, 1970), p. 168.

6. Epistemological questions are discussed frequently in the Einstein literature; see especially, John F. Kiley, Einstein and Aquinas: A Reapprochement (The Hague: Martinus Nijhoff, 1969); Niels Bohr, Atomic Physics and Human Knowledge (New York: John Wiley and Sons, 1958); and Victor F. Lanzen, "Einstein's Theory of Knowledge," in Albert Einstein: Philosopher-Scientist, vol. 2, ed. P. A. Schilpp (New York: Harper and Row, 1949), pp. 355 - 384.

7. T. S. Eliot, Collected Poems 1909 - 1962 (New York: Harcourt, Brace & World, Inc., 1963).

8. Rudolf Arnheim, "A Stricture on Space and Time," Critical Inquiry 4, no. 4 (Summer 1978): 654 - 655.

9. Evelyn Torton Beck, "Franz Kafka and Else Lasker-Schuler: Alienation and Exile — A Psychocultural Comparison," Perspectives: Perspectives on Contemporary Literature 1, no. 2 (November 1975): 40, quoting Kafka.

10. Camus, The Myth of Sisyphus, p. 21.

11. Franz Kafka, The Trial, trans. Willa and Edwin Muir (New York: Schocken Books, 1970), p. 50.

12. Franz Kafka, The Castle, trans. Willa and Edwin Muir (New York: Alfred A. Knopf, 1969), p. 24.

13. Ibid., pp. 93 - 94.

14. Franz Kuna, Kafka: Literature As Corrective Punishment (London: Paul Elek, 1974), p. 32.

15. Albert Camus, The Stranger, trans. Stuart Gilbert (New York: Vintage Books, 1946), p. 1.

16. Ibid., p. 154.

17. Jean-Paul Sartre, Nausea, trans. Lloyd Alexander (New York: New Directions, 1964), p. 160.

18. Ibid., p. 180.

19. Ibid., p. 137.

20. Samuel Beckett, The Unnamable, in his Three Novels: Molloy, Malone Dies, The Unnamable (New York: Grove Press, 1965), p. 291.

21. Ibid., p. 307.

22. George Wellwarth, The Theater of Protest and Paradox: Developments in the Avant-Garde Drama (New York: New York University Press, 1967), p. 42.

23. Beckett, The Unnamable, p. 414.

24. James B. Conant, "The Changing Scientific Scene 1900 - 1950," in The Limits of Language, ed. Walker Gibson (New York: Hill and Wang, 1966), pp. 21 - 22, quoting P. W. Bridgman.

25. Alan J. Friedman, "Contemporary American Physics Fiction," American Journal of Physics, 47, no. 5 (May 1979): 392 - 395.

26. Tony Tanner, "The American Novelist as Entropologist," London Magazine, October 1970, p. 5.

27. Lawrence Durrell, Balthazar (New York: Pocket Books, 1967), Note. See also Alfred M. Bork, "Durrell and Relativity," The Centennial Review of Arts and Sciences, 7, no. 2 (Spring 1963): 191 - 203.

28.   Lawrence Durrell, Clea, (New York: E. P. Dutton and Co., 1967), pp. 72 and 176.

29.   Ibid., p. 113.

30.   Lawrence Durrell, A Key to Modern British Poetry (Norman: University of Oklahoma Press, 1970), p. 30.

31.   Stanislaw Lem, The Investigation, trans. Adele Milch (New York: Seabury Press, 1974), p. 24.

32.   Ibid., p. 212.

33.   Ibid., pp. 204 and 205.

34.   Phillip Frank, Einstein: His Life and Times, trans. George Rosen: ed., rev. Shuichi Kusaka (New York: Alfred A. Knopf, 1963), p. 211. Einstein, of course, did not accept the gambler hypothesis; as he declares in a 1944 letter to Max Born: "You believe in the god who plays dice, and I in complete law and order. The Born-Einstein Letters, (New York: Walker and Co., 1971), p. 149.

35.   Arlen J. Hansen, "The Dice of God: Einstein, Heisenberg, and Robert Coover," Novel, 10, no. 1 (Fall 1976): 58.

15.

A Search for Form: Einstein and the Poetry of Louis Zukofsky and William Carlos Williams

STEPHEN R. MANDELL

Many American modernist poets, like their eighteenth-
and nineteenth-century counterparts, faced a revolution of
sensibility caused by science, for just as Newton had done
almost two hundred years before, Einstein had formulated a
paradigm that seemingly upset the universe. One group of
poets, in particular, sought to connect with the new science,
and to the extent that they did, they resembled those
eighteenth-century poets who readily embraced Newton's
Opticks as a source of evidence and truth. Choosing a name
that epitomized their philosophic position, they called
themselves the "objectivists." The two most important
figures in this effort were William Carlos Williams and his
close friend Louis Zukofsky.

By accepting the new science as the perceptual basis of
their poetry, Zukofsky and Williams assumed a difficult
task. Whereas Newton's view of the universe defined things
in fixed linear relation to one another, Einstein's theory of
relativity showed that time, distance, and mass vary according to an object's velocity. Both Zukofsky and Williams
agreed that poetic constructs that fixed experience into
frozen contexts would have to be replaced by those that por-
tray it as being dynamic. Only by doing this could they hope
to encompass the shifting perspectives they saw as inherent
to twentieth century phenomena. The early works of both men
are literally experiments in which they attempt to put these
ideas into practice.

As a sampling of titles indicates, Zukofsky's first
poems were not strikingly original: "Dawn After Storm,"
"Youth," "The Faun Sees," and "The Sea-Nymph's Prayer to
Okeonos." At the same time he was writing these poems,
Zukofsky was also having mixed success with free verse. However, most of the work he was producing in 1926 he chose not
to reprint in his collected short poems. Then in 1930, he
translated into English Anton Reiser's biography of Albert
Einstein which contained a twenty-four page section dealing
with space-time. Later the same year, he began section six
of his long poem "A" in which he associated poetry, music,
and physics.[1]

Throughout the 1930s, Zukofsky tried to incorporate into
his poetry the mystery and numerical integrity of nature.
Among the twelve thousand pages of his papers housed at the

Humanities Research Center of the University of Texas at
Austin, are drafts of "'A'-8" and "'A'-9." Notes attached to
these manuscripts indicate that the ending of part eight and
all of part nine were organized according to a calculus formula describing a conic section. By counting the recurrent
"n" and "r" sounds of both poems, a mathematically precise
equation can be arrived at:

> The analogy to the calculus is that the ratio of
> two sounds (r,n) is equal to the ratio of the
> accelerations of the coordinates (x,y) of a
> particle moving in a circular path for nine sym-
> metrically located points on the path.[2]

Evidently Zukofsky began experimenting in this vein before
completing "'A'-8," and finishing the poem with the aid of
the formula, he went back to the early drafts of it and
attempted to analyze their "n" and "r" patterns on the basis
of the equation. After developing his system further, he
wrote "'A'-9" entirely by formula.

"'A'-9" illustrates the use Zukofsky made of science and
how far he had come in the years since "The." Making his aim
explicit, Zukofsky attached an addendum to a draft of "'A'-9"
so there would be no doubt about his motives for composing it
in the manner he did. The numerical diversion of the poem
was, he said, a metaphor "not mathematics however prompted by
it"[3] The first stanza of the poem will serve to illustrate:

> An impulse to action sings of a semblance
> Of things related as equated values.
> The measure of all use is time congealed labor
> In which abstraction things keep no resemblance
> To goods created; integrated all hues
> Hide their natural use to one or one's neighbor.
> So that were the things words they could say:
>   Light is
> Like night is like us when we meet our mentors
> Use hardly enters into their exchanges,
> Bought to be sold things, our value arranges;
> We flee people who made us a right is
> Whose sight is quick to choose us as frequenters.
> But see our centers do not show the changes
> Of human labor our value estranges.[4]

For Zukofsky, "'A'-9" was a metaphor capable of express-
ing the subtlety and movement of the relative universe.
Within the two canzones that make up "'A'-9," words, like
iron particles in a magnetic field, arrange themselves
according to a complex system of attractions and repulsions.
Not only is there a precise series of end rhymes here, there
is also an interplay of internal rhyming (related-equated,
created-interrelated). Patterns can be discerned but none
seems to define either the stanza or the entire poem. The
stanza, part of a canzone — itself a highly complex form —
is laced with innumerable internal patterns, not the least of
which is the mathematical formula buried within its
syllables. Complicating matters further, words introduced in
the first stanza crop up with almost statistical certainty in
each of the following stanzas and in the second canzone.

This complexity places maximum stress on each word of the poem. Because of the interplay of word against word meaning becomes a function of constantly changing contexts. By loading language in this way, Zukofsky pushes each word to its denotative and connotative limits. Under the weight of this strain, syntax seems unable to hold the poet's words in check, and as a result, they appear to float free, combining and recombining in the field of possibilities established by the poem.

Like his friend Louis Zukofsky, William Carlos Williams was convinced that changes as radical as the ones occurring in science were necessary for poetry. Perhaps his earliest statement of these concerns was in his 1916 poem "To a Solitary Disciple." In it Williams addressed a lone disciple and instructs him in the proper way of observing a moon lit church steeple. At first he passively describes the steeple outlined against the sky, and then becoming actively involved in the scene, he points out the lines of force radiating from the steeple outward to infinity. Transformed, the church becomes a flower affirming the active and dynamic nature of experience:

> See how the converging lines
> of the hexagonal spire
> escape upward —
> receding, dividing!
> — sepals
> that guard and contain
> the flower![5]

Initially affected by cubism, then later by simultaneism, Williams had been trying to break free of linearity since his early work The Wanderer. Cubism, especially Duchamp's "Nude Descending a Staircase," offered him a visual model of multiplicity which he attempted to adapt to poetry in his 1920 Kora in Hell: Improvisations. Simultaneism or the simultaneous representation of different forms seen from different points of view also impressed Williams. Starting in 1917 he experimented by writing, without editing, a verbal description of images his imagination had constructed. Some of these impressions eventually appeared as Kora in Hell. By presenting the reader with a kaleidoscope of nonlogically connected sense impressions, and by combining thoughts of several people within an instant of time, Williams was able to break free of the normal sequence of experiential time.[6] Anticipating his 1948 essay "The Poem as a Field of Action" in which he connects Einstein, relativity, and poetry, Williams defined Kora in Hell as a field which "draws many broken things into a dance giving them thus a full being." He goes on to say: "The stream of things having composed itself into wiry strands that move in one fixed direction, the poet in desperation turns at right angles and cuts across current with startling results to his hang dog mood."[7]

By 1923 in Spring and All, Williams had developed a poetic form that enabled him to express the "composition by field" he discussed in Kora in Hell. Where Zukofsky overloaded language, Williams achieved similar results by stripping it down to its essentials. By doing so, he reduced the density (but not the complexity) of his poems and concentrat-

ed on the fields that individual words generate. Instead of
standing in fixed relation to one another, words dance
through a multiplicity of shifting forces and seem to flow
together like leaves in a stream or particles in a magnetic
field.   "The Red Wheelbarrow," number twenty-two of Spring
and All, illustrates Williams' technique:

So much depends
upon

a red wheel
barrow

glazed with rain
water

beside the white
chickens[8]

This poem can best be described as being analogous to a
condensed equation.   Where "'A'-9" was intricate and con-
voluted, "Red Wheelbarrow" is seemingly simple, having elimi-
nated all unnecessary elements, using only the minimum number
of words needed for communication.   But nonetheless there are
suggestions of pattern.   Words appear on the page either
singly or in triplicate and thus isolated, are themselves the
"subject" of the poem.   "A red wheel" stands apart from
"barrow," and "beside the white" is separated from
"chickens."   The colors red and white dominate and are seemingly connected by the glaze of rain water.   And of course
there is the enticing opening statement that "so much depends
upon" what is to follow.   Hugh Kenner's description of
William's use of language applies here: these are words
"dissociated into their molecules."[9]   Resembling a visual and
aural mobile, this poem affirms the natural order: its
verbal units fall into place, shift, and regroup forming a
system of ceaselessly interacting parts.
Although the objectivist project was a public relations
ploy, it did allow Zukofsky and Williams to express a set of
common ideas about poetry, and in this sense, it was (to
borrow Zukofsky's metaphor) a lens that brought things to a
focus.   But when placed in the perspective of both men's
life-long concerns, it does have a wider meaning.   As J.
Hillis Miller says, it expressed

a subtle theory of poetry which rejects both the
mirror and the lamp, both the classical theory of
art as imitation, and the romantic theory of art
as transformation.   In their place is proposed a
new objectivist art in which a poem is "Not
prophesy!   Not prophesy!/but the thing itself."[10]

To Williams the poem was "made of things — on a field."[11]   To
Zukofsky, the poem was "a context associated with musical
shape" that was composed of words "more variable than variables."[12]
In a 1969 interview, Zukofsky again reinforced this
position when he said, "The objectivist . . . as I define him
. . . is interested in living with things as they exist, and as

a 'wordsman' he is a craftsman who puts words together into an object."[13] Throughout their subsequent careers Williams and Zukofsky continued to experiment with poetic form, Williams going on to develop the relative measure and the stepped-down line, and, in 1948, writing "The Poem as a Field of Action," in which he proposed a field theory of poetic composition. Zukofsky eventually pushed beyond physics altogether into the realm of musical form, finally completing in 1975 his twenty-four-book epic that "A" had begun in 1927. For both poets, however, the "poem as object" remained a central tenet, and in this way, their early scientism provided the vital impulse for a body of work that the next generation of poets acknowledged as central to poetry in the second half of the twentieth century.

Notes

1. M. Weaver, William Carlos Williams: The American Background (Cambridge: University Press, 1971), pp. 65 - 7.
2. Marcella Spann, "An Analytical and Descriptive Catalogue of the Manuscripts and Letters in the Louis Zukofsky Collection at the University of Texas at Austin, "Ph.D. diss., University of Texas at Austin,1969, pp. 61 - 62.
3. Ibid. p. 247.
4. Louis Zukofsky, "'A'-9", 'A'-1-12, p. 112.
5. William Carlos Williams, The Collected Earlier Poems (New York: New Directions, 1951), p. 167.
6. Bram Dijkstra, The Hieroglyphics of a New Speech: Cubism, Stieglitz, and the Early Poetry of William Carlos Williams (New Jersey: Princeton University, 1969), pp. 68 - 69.
7. William Carlos Williams, Selected Essays (New York: New Directions, 1954), p. 15.
8. William Carlos Williams, The Collected Earlier Poems of William Carlos Williams (New York: New Directions, 1938).
9. Hugh Kenner, A Homemade World (New York: Knopf, 1975), p. 59.
10. J. Hillis Miller, Poets of Reality: Six Twentieth-Century Writers (New York: Altheneum, 1974), pp. 309 - 10.
11. William Carlos Williams, The Autobiography of William Carlos Williams (New York: New Directions, 1951), p. 333.
12. Louis Zukofsky, Prepositions (New York: Horizon, 1968), p. 24.
13. L. S. Dembo, "The 'Objectivist' Poet, Four Interviews," Contemporary Literature 10 (Spring 1969): 203.

Part VII

Politics

16.

Einstein on War and Peace

THOMAS RENNA

The historiography of peace literature has ignored Albert Einstein. Historians sympathetic to Einstein's pacifism are often embarrassed by his fervor for world government and his support of atomic bomb research at the start of World War II. Those who are less pacifist gleefully point to Einstein's abandonment of pacifism in 1933. Scholars smile tolerantly at his naive "idealism." Critiques of Einstein's views on war tend to be summaries of his written and oral statements.[1] But Einstein's unique role in the Western peace tradition cannot be grasped until his ideas are placed in their historical context.

It is obvious that Einstein's views on the causes of war were similar to those of other scientists and writers during the 1920s and 1930s. As did most peace-minded intellectuals of the interwar era, Einstein blamed nationalism for the world's ills.[2] The very existence of armies and armaments served to inculcate nationalist values.[3] So too, competition in the economic and political spheres fostered destructive conflicts within and between countries.

Thus Einstein generally shared the common assumptions of at least some writers concerning the basic causes of modern war. He suggested five ways to eliminate war. First, nation states must become politically democratic.[4] Only an enlightened populace can offset the corrupting influence of militarists and businessmen. In a democracy it is harder for the government to enslave its people and lead them into senseless wars.

Second, education must be reformed. People should be made aware of their responsibilities as world citizens.[5] History, geography, and economics are to be taught from an internationalist viewpoint.

Third, intellectuals must organize and pressure political leaders to stem the arms race.[6] The new elite must warn of the risks of war, and preach the advantages of peace. Bertrand Russell was the most articulate spokesman for pacifism as a public policy.[7] Einstein preferred to urge individuals to refuse both military service and work in defense industries.[8]

Fourth, science and industrialism are to be accepted as positive goods. Post-World War I writers emphasize the moral neutrality of science. The issue was no longer the goodness or badness of science as such, but the intentions of those who controlled it.[9] Science must remain free and independent.

Fifth, conflicts must be resolved by international arbitration. The chief obstacle to peace was national sovereignty. Einstein and other pacifists wanted the League of Nations to assume the status of a world government with the legal sanctions to enforce the peace.

It would seem, then, that Einstein's war posture during the period 1919 to 1939 was hardly unusual among many educated Europeans and Americans. But why did he abandon his pacifism so quickly after Hitler came to power in 1933?[10] He was, to be sure, consistent in his war/peace philosophy before and after 1933, as Einstein himself always insisted.[11] But it is misleading to classify him together with all pacifist intellectuals prior to 1933. A more careful look at the content and style of his public statements reveals consistency on a different level.

Einstein emphasized the aggressive instincts of human beings. In this assessment of man he is perhaps closer to Freud's pessimism than to Russell's rationalism and Julian Huxley's evolutionary humanism.[12] Unlike many of the idealistic pacifists of the 1930s Einstein concluded that debased human nature was precisely why a strong world government was needed.[13] He thought that the Communists were mistaken in their belief that institutional change would bring equality and cooperation. Whereas leading pacifists such as Aldous Huxley stressed economic forces, sociological patterns, and improved technology in their analyses of war, Einstein preferred to look at personalities. Germany was headed for war because the gang of thugs in power was determined to plunge Europe into chaos.[14] Compare this simplistic critique of Nazi Germany with that of H. G. Wells — whose universalist outlook Einstein admired — who considered National Socialism as a symptom of a worldwide malaise caused by technology and competition.[15]

This comparison with H. G. Wells is instructive. Wells sprinkled his prose with wit, sarcasm, and allusions to current events. The humorless Einstein preached repentance and Armageddon. Einstein's remarks on peace were usually brief and repetitious. He issued warnings, not action plans. Einstein's call for world government was not utopian or idealistic, as is often said. For him the issue was survival, not earthly bliss. World government was an international police force aimed at stemming man's propensity to violence; it was not, as for Wells, the beginning of a world revolution that would usher in the millennium of love and joy. Wells' language follows the utopian literary tradition. Einstein's blasts against European materialism and blind conformism stand closer to the apocalyptic legacy.[16] Man was locked in the ongoing struggle between good and evil, with the latter always salient. World government would not end the struggle; it would only make it more difficult for demonic persons to start wars. Wells' view of good government lies in the Aristotelian tradition; Einstein's, in the Augustinian. Einstein was incapable of compromising on the question of supranational authority. Only the reprobate could fail to see the inherent truth in such a concept.

The doomsday tone of Einstein's peace is also evident in

his discussions of the peacemaking function of the intellectual in general, the scientist in particular. Einstein had
little faith in organizations. He envisioned groups of
intellectuals as aggregates of individuals who would prod
political leaders to outlaw war, and press for an effective
League of Nations. He was not very hopeful that such organizations could actually influence the course of events, but he
felt nevertheless that such efforts were necessary. His support of nuclear scientists after World War II was consistent
with his pleas for the internationalization of science before
the War. Scientists in their capacity as citizens were obli-
gated to give witness to the goals of freedom and justice.
Einstein's own role in organizations, such as the Committee
on Intellectual Cooperation, was always passive. While he
gladly lent his name to virtually any group which had peace
in its title, he never actively participated in its operation. While Russell sat in jail, Einstein signed manifes-
toes. He was the outsider whose imprimatur added moral
weight to newly founded pressure groups.

Einstein's ambivalence about social commitment was typical of the crisis of interwar intellectuals. He was a hold-
out for the universalist outlook portrayed in Julien Benda's
Betrayal of the Intellectuals (1928). With Ortega y Gasset
he lamented the revolt of the masses. With Jose Oswald
Spengler he saw the decline of the West's heritage of freedom, individualism, and humanitarianism. Just as Thomas Mann
groped for a new social role for the intellectual in The
Magic Mountain, Einstein groped for a meaningful social role
for the scientist. His solution was pacifism. The scientist
should not, to be sure, succumb to the rush to engage; he
should remain faithful to his research and the international
pooling of scientific knowledge. But the scientist as
citizen must work for a world government if for no other
reason than to maintain the freedom of scientific research.

The world's most famous scientist said surprisingly
little about the value of applied science. While his enthu-
siastic colleagues might see in science the panacea for all
social problems, Einstein spoke more about the possible mis-
use of science than its benefits.[17] Science possessed no
moral value; it was an instrument for either good or evil.
To ensure that science be used for worthy ends, democracies
— better yet, socialistic democracies — should be created
throughout the world. When scientists are allowed to communicate freely across national borders, the outbreak of hosti-
lities becomes less likely. Science is virtually the only
international force which remained after the catastrophe of
the Great War. While his fellow scientists stressed the
social return from applied science, Einstein made a more
subtle point: the scientist's act of research was a way of
maintaining universal human values. In the solitude of its
laboratories the scientific elite was doing more to preserve
the integrity of the individual than were activist intellectuals with their mania for social causes.[18] Nationalist wars
are, after all, precipitated by those who have lost contact
with their own humanity. Thus, from the point of view of
world peace, basic research is indispensable. While he
applauded the efforts of science to diminish the indirect
causes of war (ignorance, poverty, disease), Einstein did not
give much attention to them.[19] Strange to say, Einstein's

somewhat esoteric view of the salvific effects of pure
science resemble the search for identity of a contemporary
monk, Thomas Merton.[20] Einstein looked to science as the
last refuge of a disintegrating Western culture. Science was
the sole point of reference in an unstable and illusory
world.

Einstein's notion of science and war cannot be easily
aligned with the intellectual trends of the interwar years.
Thus he wrote eloquently of the individual's dignity and the
sublimity of the creative act. But he just as often down-
graded reason and the potential for human growth. The individual counted as nothing next to the totality of humankind.[21] Submission to the state was demeaning; to the world
community, spiritually uplifting. His views on the educa-
bility of the working classes fluctuated.[22] So too, he was
unsure about the ability of intellectuals to teach peace.

It would be, therefore, historically inaccurate to
impose a theory over these scattered thoughts. They must be
accepted as they stand. There were no easy answers in that
uprooted age. Einstein's inconsistencies are themsleves
illustrative of Europe's confusion.

Einstein's conceptions of education and democracy were
generally in step with those who believed peace could be
attained through mass education within democratic nations.[23]
Einstein wanted the schools to produce an elite steeped in
universalist values, and a public sensible enough to select
pacific leaders.[24] Einstein understood little of the give-
and-take of the democratic process. During the 1920s he
romanticized American democracy as a viable alternative to
totalitarianism. But his faith waned as the American government hesitated to challenge Nazi Germany, bullied nuclear
scientists after 1944, and behaved selfishly vis-à-vis the
USSR and the United Nations. From 1919 to 1955 Einstein was
the West's conscience. In this role he followed the tradition of the detached wise man who descended from the mountain
to correct his wayward people. Einstein's focus on personalities instead of institutions is part of this age-old
pattern. Personal example counted for more than social reform.[25] His call for world peace and order continues the
tradition created by the ancient Hebrew Prophets, Hellenistic
philosophers, early medieval monks, Erasmian humanists,
eighteenth-century philosophes, and nineteenth-century lit-
terateurs. Einstein appeared at a time when Europe had lost
confidence in its traditional leaders. After 1919 Europe
made as its culture hero not a soldier, litteratus, or busi-
nessman, but rather an offbeat physicist. The iconoclastic
1920s had smashed all its idols. Cynicism and fatigue prevented the idealization of a political figure. While most
Europeans rushed to join a group and adhere to an ideology,
they also needed a counterhero who was unaffected by these
attachments. Einstein was the refreshing nonconformist, a
human being among fanatics, an individual among faceless
hordes, the enigmatic holy man with all the answers. This
serene saint kept aloof from the noise, trivia, dogma, and
activism of a shattered Europe. The silly Einstein stories
in circulation from 1919 to 1939 were intended to prop up
this image — an image that Einstein did little to destroy.

Einstein's enormous popularity as a social critic was
the culmination of a tradition that started in the

seventeenth century with the rise of experimental science.[26] Francis Bacon suggested a new scientific elite that would ensure lasting peace.[27]   Descartes' *esprit geometrique* was crudely applied to principles of government and social structures.[28]   Some philosophes, such as Helvetius, extended the scientific method to education; they used natural law to criticize "unnatural" privilege.   In the nineteenth century most utopian and socialist writers, although not usually scientists, welcomed the advance of science and technology. From 1850 to 1914 the scientists emerged as a distinct group.   Following World War I some scientists attacked their colleagues for assisting the military, and abandoning the legacy of international science.

Thus, Einstein's life as peace prophet falls into three periods: (1) 1919 – 33:   for Einstein, absolute good is pacifism and absolute evil is nationalism/militarism; (2) 1933 – 45:   good is world government (to preserve civilization and freedom), evil is Nazi Germany;[29] and (3) 1945 – 55: world government/control of nuclear weapons is the moral imperative.

Einstein's attitudes toward war and peace — so simple yet so profound — reflect the realities of the modern age.   He realized that henceforth war and science would be intertwined.   He foresaw the decline of the just war concept, and the need for a redefinition of pacifism for the nuclear age.   Traditionally the role of the peace witness was to demand personal reform and the return to harmony with divine or natural law.   With Einstein this tradition comes to a close.   He was the last prophet.

Notes

1.   See V. Hinshaw, "Einstein's Social Philosophy," in *Albert Einstein: Philosopher-Scientist*, ed. P. Schilpp (Evanston, Ill.: Library of Living Philosophers, 1949), pp. 649 – 61; A. French, "Einstein and World Affairs," in *Einstein: A Centenary Volume*, ed. A. French (Cambridge, Mass.:   Harvard University Press, 1979), pp. 185 – 97; P. Simon, "From Pacifism to the Bomb," *Einstein*, eds. L. Broglie et al. (New York 1979), pp. 151 – 71.

2.   O. Nathan and H. Norden, eds., *Einstein on Peace* (New York: Schocken Books, 1968), pp. 35, 38, 44, 152, and 163.

3.   *On Peace*, p. 113f, p. 241f.

4.   *On Peace*, pp. 209, 284, and 312 – 314.

5.   *On Peace*, pp. 144 – 46, 174, 253, 273f; A. Lief, ed., *Albert Einstein: The Fight Against War* (New York: The John Day Co., 1933), 38; C. Seelig, ed., *Ideas and Opinions* (London: Souvenir Press, 1973), pp. 62 – 74.

6.   *The Fight Against War*, p. 43; *On Peace*, pp. 140 – 42; *Ideas and Opinions*, p. 91f.; *Out of my Later Years* (New York: Philosophical Library, 1950), pp. 152 – 55.

7.   See especially Bertrand Russell's *Which Way to Peace?* (London 1936).

8.   Einstein was a leading spokesman for individual war resistance from 1928 to 1933.

9.   *On Peace*, pp. 94, 104f., 312.

10.   After 1933 Einstein defined himself as a "realistic pacifist":   one who works for both the defeat of the Nazis

and the creation of a supranational government. See On Peace, p. 276, passim.

11. Cf. P. Frank's quaint explanation of Einstein's inconsistency on the basis of his scientific principles; Einstein (New York: A. A. Knopf, 1947, p. 245f.

12. On Peace, pp. 196 - 202. Cf. Freud, Civilization and its Discontents (New York: J. Cape & H. Smith, 1930); P. Michelmore, Einstein: Profile of the Man (New York: Dodd, Mead, 1962), chapter 9.

13. On Peace, pp. 236 and 242.

14. On Peace, pp. 219 - 22.

15. H. G. Wells, The New World Order (New York:A. A. Knopf, 1940); The Common Sense of War and Peace (Middlesex: Penguin Books, 1940).

16. On Peace, pp. 73, 161, 179, 262 - 83, and 330.

17. On Peace, pp. 218, 269f., 283f., 330f, 355f., 428f., and 493f.; Ideas and Opinions, pp. 219 - 22, 326 - 28.

18. On Peace, pp. 534 - 37.

19. Out of My Later Years, p. 123f. Cf. Julian Huxley, "Peace through Science," in Challenge to Death, ed. P. Baker et al. (London 1934), pp. 287 - 304; K. Compton, "Man and Technology," in Science and Man, ed. R. Anshen (New York: Harcourt, Brace and Co., 1942), pp. 309 - 24.

20. Merton first hints at the social effects of interior solitude in his autobiography, Elected Silence (London: Hollis and Carter, 1949), a theme which he integrates into his pacifism in his post-1955 writings.

21. On Peace, p. 95.

22. For example, On Peace, p. 315f. After 1935 Einstein steadily lost faith in the possibility of intellectuals influencing either the general public or political leaders in the cause of peace, although he never stopped trying.

23. C. Howlett, Troubled Philosopher: John Dewey and the Struggle for World Peace (Port Washington, N.Y.: Kennikat Press, 1977), Chapters 4 - 7.

24. On Peace, pp. 144 - 46, 209, 284, and 429.

25. On Peace, pp. 274, and 284.

26. The history of scientists as social critics has yet to be written, but see J. Dupre and S. Kakoff, Science and The Nation (Englewood Cliffs, N.J.: Prentice-Hall, 1962); J. Bernal, The Social Function of Science (Cambridge, Mass.: M.I.T. Press, 1967); H. and S. Rose, Science and Society (London: Allen Lane, 1969); K. Silvert, ed., The Social Reality of Scientific Myth (New York: American University Field Staff, 1969), Chapter 2; J. Ben-David, The Scientist's Role in Society (Englewood Cliffs, N.J.: Prentice-Hall, 1971), Chapter 4 - 7; D. Schroeer, Physics and its Fifth Dimension: Society (Reading, Mass.: Addison-Wesley Publishing Co., 1972).

27. Bacon's New Atlantis began a tradition which persisted in utopian literature down to 1914; cf. Saint-Simon and Lytton.

28. A.G.R. Smith, Science and Society (London 1972), chapter 4.

29. At first Einstein cursed Hitler and his associates. But during the war he came to believe that the German people were hopelessly warlike. Cf. On Peace, p. 366; R. Clark, Einstein: The Life and Times (New York: World

Publishing Co., 1971), p. 492.   Most pacifist intellectuals
— Wells, Russell, Huxley — deemphasized the differences
between Germans and other peoples.   The Einsteinian style,
however, preferred hyperbolic contrast.

17.

Political Origins and Significance of China's Einstein Centennial

EDWARD FRIEDMAN

A special celebration took place on behalf of Albert Einstein on February 20, 1979, in Beijing, China. Over 1,000 Chinese scientists attended a commemorative meeting with foreign dignitaries in attendance. At the meeting, Dr. Zhou Peiyuan, the acting head of China's Scientific and Technical Association, spoke. Zhou, as a physics graduate student at Columbia University, had traveled to Princeton to learn from Einstein and apparently had been, with Guo Muoruo, Mao's closest adviser on science. In his talk, Dr. Zhou stated that Einstein had recently become a political issue in China:

> This universally esteemed scientist met with humiliation and slander in our country during the Lin Biao - Gang of Four period. It is with a view to restoring in China the honored position of this great scientist that this meeting in his honor is being held today.[1]

I want to suggest why the meaning in China of Einstein's work became a matter struggled over among China's ruling groups.

Most likely, Mao Zedong's views of Einstein's work were set by the end of the extended May Fourth period, that extraordinary era in Chinese history starting during World War I when all ideas could be scrutinized and totally rejected and new, strong commitments to modernize China were made. Dominant tendencies of that era included the rejection of liberal mechanical theories, a turn to indigenous wisdom, a desire for something newer and better. All of these would facilitate an extraordinary May Fourth reception for Albert Einstein. China was flooded with articles about the "Revolution in the Scientific World." Bertrand Russell, in visiting China in 1920, would link Einstein and Lenin as the two greats of the age.[2]

The events, the timing, and the almost universal desire of modern people to want to seem scientific suggest that Mao Zedong and others of his generation would be defining and linking their politics with their own understanding of Einstein. One suggestive piece of evidence for this early linkage is the lack of any subsequent interest by Mao in Engels' main writings on science (not published in any language until 1925) or in the demeaning Soviet Stalinist debates on Einstein, relativity, and quantum theory.[3] Mao apparently already knew what this Einsteinian revolution implied.

Among the ideational forces which stirred the winds in China that blew away Newton's mechanics, the key event was the First World War, often referred to in China as the European Civil War. That massive bloodletting among the advanced powers gave the lie to European commercial liberal optimism.[4] The liberal truth had been that commerce was civilization, that trading relations were replacing warrior relations, civil society replacing feudal-military societies. As a result, ever since the end of the Napoleonic wars, the expanding European civilized world was experiencing peace and progress. To be sure, as with Britain's invasion of China known as the Opium War, the advanced supposedly had to drag backward people kicking and screaming into the orbit of progressive civilization, but once in, the military imperative naturally gave way to the mutual benefits of generally expanded commercial wealth. This sketch of progress was covered over with the blood of advanced Europeans in 1914. A whole world view was thereby called into question. By 1919 any educated Chinese informed about European affairs had to suspect that the supposed natural link between liberalism and its scientific promise of ceaseless advance toward a reasonable world at one with truly human potentials had been weakened, if not snapped.

Chinese were correctly informed that leading Western intellectuals — Bergson, Kropotkin, Russell, Dewey, James, Euchen, Einstein — were now looking to Asian thought, to traditional Chinese and Indian wisdom, for a corrective to the amoral, murderous tendencies of the liberal West. A new and better world would have to build on a synthesis of the new Einsteinian science which incorporated the old social wisdom of such as China. Given such an intellectual milieu, it is not surprising that Mao in 1973 would inform Nobel Prize - winning physicist, C. N. Yang that on these ultimate questions of the divisibility of matter the ancient Chinese philosopher Xun Ci had much to say.[5] Seeking harmonies between the new physics and Chinese wisdom was not merely an artifact of Mao's nationalism. It was also the preferred mode of thinking of the great, new Western physicists.

In China — and other countries suffering the indignities of foreign domination — accepting the Leninist interpretation of the First World War as an intraimperialist struggle opened the way to seeing the war as a cause for optimism for colonial and semicolonial people. What weakened and bloodied the West made it possible for people of the East to rise up and win their independence.[6]

One clue to this difference in outlook between China and the West is in the reception of Henri Bergson's Einsteinianism. Bergson, an acquaintance of Einstein, developed a relativistic notion of reality in which Newtonian determinism was supplanted by subjective intuition. Bergson was invited to China. That Bergson's world view ended a "fog of pessimism," not only made him attractive to Chinese revolutionaries, but also made him increasingly silly in the increasingly pessimistic West. His hopeful world view centering on human creativity became in the West just another casualty of the

war.[7] A primary way for comprehending the new realities for
populist Chinese was precisely the one way the Western culture of that era would no longer permit.

There is no difficulty in showing that much in Bergson's
1911 book, Creative Evolution, which was translated into
Chinese, is in harmony with much of what eventually became
known as Maoism. For example, the harmonies between
Bergson's and Mao's biological metaphors of change, in which
metabolism and regeneration are identified, is extraordinary. Moreover, it would have been natural in the atmos-
phere of the time for Mao to seek progressively liberating
energies in China's people rather than in large, bureaucratic
mechanical organizations.

Although we have no concrete reflections of Mao on the
new physics until after the People's Republic was established
in 1949, we do have some indications from the 1930s that Mao
already knew that he and his revolution were at one with the
new truth. There is more than harmony with Marxist ideas of
revolutionary qualitative change. Meanwhile, in both 1931
and 1937, Einstein spoke out against the Japanese invasion of
China, aligning himself with the forces of China's independence struggle.[8]

We probably should not make too much of Mao's 1930
statement about revolution "advancing in a series of waves."[9]
 The Comintern had long described revolution in terms of the
ebb and flow of tides and waves. But Mao had this as part of
a general conception of change in which progress simply could
not be continuous or a matter of inertia in which events
"follow a straight line" since they actually move by "twists
and turns."[10] But at a deeper level Mao knew, apparently
with Werner Heisenberg and his Einsteinianism, that one
learned about the world in the experience of changing the
world. Formulas could not be applied from the outside in the
Newtonian way. This is summed up in the famous phrase from
Mao's lecture "On Practice" that "If you want to know the
taste of a pear, you change the pear by eating it yourself."
Usually Mao is treated either as a homey philosopher or a
country bumpkin and scientific ignoramus. In fact, Mao was
a modern thinker asserting his Einsteinianism. To learn
where the electron is you must move it. "If you want to know
the structure and properties of the atom, you must make physical and chemical experiments to change the state of the
atom."[11]

Mao thus put down his Party adversaries who argued from
the writings of Marx and Lenin or from the experience of the
Bolshevik revolution. The ultimate grounding of Mao's revolutionary practice instead was that he, not his opponents,
was in harmony with Einsteinian science. Quite naturally,
therefore, before his two subsequent revolutionary thrusts,
the break with the Soviet model in the 1950s and the Cultural
Revolution in the 1960s, Mao returned for inspiration and reassurance to this ultimate source of wisdom.

Once in power, knowing what he believed, Mao avoided
Stalinist polemics, and looked for and found what was con-
genial to his Einsteinianism in the great Japanese nuclear
physicist Sakata Shoichi. Mao would stick with Sakata for
the rest of his life. Intriguingly, at the same moment that
Mao was turning to Sakata, Sakata was reading Mao's philosophical essays. Sakata saw in Mao what Mao saw in Sakata.

Their Einsteinianisms, their conceptions of the new physics, were the same.

Early in the 1950s Mao Zedong had his attention called to Sakata's work which Mao proceeded to delve into. Mao dispatched Guo Muoruo and Liao Zhengzhi to Japan in 1955 where they met Sakata at Kyoto University. The next year Sakata responded to an invitation to visit China.[12] We cannot be sure which or how many of Sakata's articles Mao went over, but Sakata's major pieces of that period promoted a view that would intrigue Mao and they were packaged not only in a language that Mao would appreciate — practice, materialism, revolution, Marx, etc. — but in metaphors which were Mao's too. As Sakata would cite Kant and Laplace and a view of nature in which nature is not a given but a process of endless birth and death,[13] so Mao would lecture those around him, generals and all, on Laplace and Kant.[14]

Sakata saw knowledge in terms of qualitatively different levels. Learning was like peeling an onion with an infinite number of layers. By resolving the "contradictions" at one level, one progressed to the contradictions at the next level. In light of "the new elementary particles ... discovered one after another almost monthly," it became necessary to search for particles within particles, different theories for separate strata.[15]

While in Beijing, Sakata, in addition to participating in intimate conversations with Guo and Dr. Zhou, was given numerous platforms from which to proclaim his philosophy of science. He spoke at Beijing University and the Academy of Science. He was interviewed by the press. His speeches were translated and broadcast over the radio. His topics were the philosophical ones that Mao preferred, composite particles and theoretical physics. Two days before he left Beijing, Sakata was presented with copies of Mao's "On Practice" and "On Contradiction," which, of course, he had already read.[16]

In later years Sakata continued working and philosophizing in ways Mao could not help but appreciate. Sakata declared that any "revolution cannot arrive at the final understanding of nature at a time and therefore it must be carried out through many steps and may be continued rather infinitely."[17]

When the little leap forward of 1955 - 1956 had produced real setbacks after apparent initial advances, Mao retorted with the popular Einsteinianism of our age:

The law of imbalance is a universal precept of development. Progress is made out of twists and turns and spirals.
Sudden change is the most basic law in the universe. The production of new matter is a good thing. We Communists hope for the change of matter, so-called leaps forward...sudden changes. The destruction of balance is a leap forward; the destruction of balance is superior to balance. Imbalance and big headaches are good things.[18]

It was natural, Mao argued, that progress should occur in waves rather than a straight line. This was a law of nature.[19]

In other words, accumulation proceeds in a wave-
like manner or in spirals. Since everything in
the world is itself a contradiction, a unity of
opposites, its movement and development is wave-
like. The light emitted by the sun is called
light waves, the waves transmitted by radio
stations are called radio waves and sound is
carried by sound waves. . . . Such is the
undulatory nature of the movement of opposites in
all things.[20]

Wave-like advance. All movement is waves. In
natural science there are sound waves and
electromagnetic waves. All movement is wave-like
advance. This is the law of the development of
motion. It objectively exists. It is not changed by human will. All our work...is wave-like
advance. It is not a rising straight line.[21]

Wave-like advance is inevitable. The advance of
economic construction is wave-like. It rises and
falls. One wave follows another. If you understand this point, this year's bold advance is
nothing surprising. Next year there will be some
contractions. That's it.[22]

Despite the philosophical defense of his Great Leap
policies in terms of the new physics, Mao never tried to re-
peat those disasterous initiatives. He did, however, rededi-
cate himself to preventing what he considered to be a new
bureaucratic class from entrenching itself in China. This
notion of a continuous revolution led Mao even more insis-
tently to call to his defense the new physics, especially
Sakata Shoichi's philosophical understanding thereof. It
supported Mao's view that the young and new in strong waves
should push out the old.[23]   Mao combined Buddhism and
Taoism, Lenin and Sakata Shoichi, and asked what to do with
the situation in China where much power supposedly was still
held by class enemies. He insisted that even in a communist
society there would be more stages, more qualitative
changes. He called attention to the continuing discovery of
particles and antiparticles.

The electron still has not been split. But the
day will come when it can be divided. . . . If
it is exhaustible then it is not science. . . .
Matter develops. It is infinite. Time and space
are infinite. Space, both at the macro and micro
level, are infinite. If it is infinite, it can
be divided. That's why scientists will still
have work to do a million years from now. I
really enjoyed Sakata Shoichi's article in the
"National Science Research Bulletin" on basic
particles. I have never before seen such an
article.[24]

On August 23, 1964 Mao met with Sakata who was in
Beijing attending a physics symposium which ran from August 2
to August 31. The next day Mao called a meeting of his top

philosophy of science people, Zhou Peiyuan, Yu Guangyuan and others to discuss the implications of Sakata's work.[25]   He went over change in the manifestations which had long fascinated him and made science itself his symbol:

> Everything in the world is [a process of] change. Physics also is [a process of] change. Newtonian mechanics is also [a process of] change. The world went from not having Newtonian mechanics to having Newtonian mechanics. And afterwards from Newtonian mechanics to relativity theory. This is dialectics itself.[26]

It is very difficult not to read this talk of Mao's in terms of the Red Guard Cultural Revolution's mass chaos, which he soon set off against his alleged enemies. Einsteinianism was its deepest legitimization.

That Mao rested his Marxism on Einsteinianism had some great virtues. Confronting the changes in world view wrought by the new physics legitimated the notion of no dogmas, not even Marxist dogmas. If life and science were ceaseless change from pre-Newtonian physics to Newtonian physics to the post-Newtonian new physics and on to something else and something else and so on, then, quite naturally, "I say that Marxism would have its birth, development, and death."[27] The future would laugh at the hauteur of the present. This notion that human praxis created new worlds requiring new explanatory schemes was worlds superior to the Marxist dogmatism which always tried to reconcile new social creations with some passage in a text of Marx from over a century before. That was conservative, religious dogmatism, a priesthood trying to preserve a faith and parading the naked old emperor while shouting huzzahs for his nonexistent eternally true garments. Mao would have none of that. His Sakata based Einsteinianism insisted on the endless need to rethink the ever-new universe of dynamic human change. Yet in his own politics, Mao would make this wonderful abstraction concrete in a way which would rip and shred Chinese society even in the world of nuclear science.

Sakata's 1964 trip to China coincided with Mao's deepening commitment to another split in the Chinese Communist Party, a need for what became the Cultural Revolution launched in 1966, and which led in 1965 to attempts at legitimizing the notion of continuing divison as truth. Mao had read, Sakata's April 1961 article, "Dialogues Concerning a New View of Elementary Particle," which was translated from the Japanese language Proceedings of the Physical Society of Japan[28] and published in mid 1965 in the theoretical journal of the Chinese Communist Party.[29] It was an extraordinary event. As far as I know, nothing like it ever happened before or since in that Leninist theoretical journal. Even more extraordinarily, it was followed up, in another issue of Red Flag, by a series of essays discussing its significance.[30] Two key political points were made.

> In human society and the natural world a unified entity always tends to split into different parts. . . . To deny the universality, the absoluteness and limitlessness of "division" is

to deny the limitless development of the natural
world and society, to deny the infinite development of understanding, to cause the dialectic to
be severed in the middle, to bring to a halt and
to destroy natural science and social science.[31]

Contradictions can only be resolved through struggle.
This is true even under social science.[31]

The Soviet Union was distinguished from Einsteinian
China by its epistemological and related practical errors on
this matter of contradiction, division, and struggle.

In his 1961 article, Sakata, in additon to offering a
series of theories for a whole host of unresolved issues in
microphysics, set out most clearly his philosophical assumptions. He took on theories of ultimate truth or limits on
knowledge in one form after another going back to ancient
Greece, from mathematical formalizations to a quest for simplicity of matter, from Niels Bohr's "mist of Copenhagen" to
Werner Heisenberg's "Urmaterie." He argued instead for a
straton approach, one in which any physical theory held true
only for a particular level and only contingently. What made
for the contingency would lead research workers to progress
to a deeper level in an infinite number of strata in which
new laws would become necessary. The revolution of knowledge
was not merely continuous; it was inherent in the nature of
reality, the point Mao insisted upon.

This legitimized for Mao a view that the new issue which
he would now address (how to prevent a capitalist restoration
in a socialist country) would require new practice and theory
for the new level of reality. That view of Mao was held to
by those most firmly committed to China's Cultural Revolution. But the Cultural Revolution itself and subsequent
Chinese politics were not guided by Mao's Einsteinianism.
What concerns us here is Mao's politicization of Chinese
nuclear science. There is a large difference between a group
of scientists following up on Sakata's lead and a state forcing all research into this mold. Chinese nuclear physicists
found themselves confronting Maoist Einsteinian imposition,
something akin to a Chinese style "Scopes trial" or "Lysenko
case."

It is worth keeping in mind just how much importance Mao
gave to Sakata's work. In July 1966, Beijing convened an
international summer physics colloquium. A number of Japanese physicists attended, but intriguingly, not Sakata.[33]
Zhou Peiyuan, Mao's favorite physicist for discussions on
these issues, headed the Chinese delegation, but Mao was very
busy at this time. On August 1, 1966 the Eleventh Plenary
Session of the Eighth Central Committee of the Chinese Communist Party met. During that session, Mao launched his
long-since philosophically presaged division, the Cultural
Revolution attack on top Party people. Where would Mao be
the evening before that all-important political event? At 10
P.M. the night before the crucial Central Committee meeting
opened, Mao met with the delegates to the physics conference
investigating the applicability of the Sakata model.[34]
In the aftermath of the economic, educational, technological, and scientific setbacks of the Cultural Revolution,
the fundamentalists trying to "continue the Revolution"

attempted to discredit Einstein.[35]   Indeed, in 1970, under
the direction of what their opponents then called "the Shanghai Gang," led by Zhang Chunquiao and Yao Wenyuan, a criticism group at Shanghai's Academy of Science was already given
the mission of attacking Einstein.   It was Yao Wenyuan who
made his own this project of developing "a proletarian class
theoretical system for the natural sciences."[36]   This faction
urged a halt on reversing the unjust verdicts reached against
scientists during the Cultural Revolution.   Their spokesman
in Beijing, Zhi Chun, paralyzed Beijing University and
Qinghua University with his threat to launch a campaign to
seek out and punish a batch of so-called rightists every two
years.   The fundamentalists called for rectifying the supposedly corrupted intellectuals involved with international
work by sending them to the villages to use their foreign
languages to criticize landlords.[37]   Actually, China had not
had landlords for a quarter of a century.

In Shanghai, Yao Wenyuan's people organized a group
calling itself Li Ke, a homonym for Academy of Science.   They
sought help from professors at Shanghai's Fudan University in
putting together material to prove that Einstein was in fact
a reactionary capitalist, that all science had a class content, that Einstein's science was antisocialist science and
that therefore the powers in Beijing associated with it were
not socialist.

From a philosophy of science standpoint, the Academy of
Science essays were full of "laughable errors."[38]   There was
nothing funny, however, about what was happening.   The goal
was to link Einstein with vilified Chinese leaders, making
Einstein and friends a priori idealists (in this case one
stressed Einstein's rationalism) on the surface who in fact
were mere empiricists (stressing the experimental support
which Einstein referred to), that is, supposedly not people
realistically motivated by the great cause of continuing the
socialist revolution to the end.

The fundamentalist's political goal was to link Einstein
and relativity with the idea that truth was relative, that
knowledge had limits, and then to dub these a counter revolutionary negation of the correct position that truth was in-
exhaustible and limitless.   But, as we have seen, that latter
position was in Mao's eyes in fact Einstein's and Sakata's.
Whatever the misleadingly brutal propaganda involved, the
campaign did result in Chinese physicists and philosophical-
ly-minded college students learning and repeating the
Einstein arguments against Bohr for being satisfied with
wave-particle dualism.

The Shanghai Academy of Science group's understanding of
Einstein, while buttressed by scholarly footnotes following
an obviously serious look into Einstein's work and commen-
taries thereon, went no higher than the popular cultural misunderstanding.   To the Academy's fundamentalists, Einstein's
relativity theory meant that one point of view was as good as
another, that everything was relative.   There was no mention
of Einstein's notion of frames of reference, which permits
one to comprehend the merely apparently relative.   The Academy of Science group was utterly scandalized by Einstein's
credo, his supposedly Spinozan confusions (which actually can
be harmonized via Kant with Marx's notion of a species
nature),[39] and by evidence that invented mathematical equa-

tions can precede physical results. The Academy group concluded that Einstein's absolutizing of the relative along
with his appeals to ultimate faith were precisely what
capitalist imperialism in decline called for. "Einstein's
world view is a concentrated reflection of this political
imperative of capitalism."[40]
    If their logic was shallow, their political target was
clear. The goal was to link Einstein and Prime Minister Zhou
Enlai and discredit both. They called Einstein "the greatest
Confucian of all western scientists.[41] They hoped to smear
Zhou as a Confucian as part of an anti-Confucian campaign
launched at this time.
    Politically, one interesting feature of the series of
articles published in Shanghai against Einstein, is that they
never became part of a political movement outside of Shanghai.[42] They began in 1973 with essays in the Fudan University Studies Journal and moved in 1974 to disquisitions in
Shanghai's Journal of Natural Dialectics. But this attack on
Einstein, as the greatest and most prestigious reactionary
capitalist academic authority, never got out of Shanghai,
never took national form, even while the Shanghai fundamentalists' anti-intellectual campaign did win some brutal and
bloody national battles. This suggests that on the issue of
Einstein and a philosophical grounding of their politics, the
fundamentalists were in conflict with Mao Zedong, who otherwise did seem to sympathize with many of their anti-intellectual policies. The fundamentalists would judge reality by
Marxist a prioris, whereas their political opponents claimed
an independent need to study natural science in its own
terms.[43] Neither side sits easily in Mao's own position
which gives priority to the new physics, properly understood,
that is, understood in terms of Einstein and Sakata.
    After the 1976 death of Mao and the arrest of the lead-
ers of the fundamentalists, the Einstein who rose to pro-
minence at China's large and happy celebration of the
Einstein Centennial Year was not the Mao-Sakata version of
the new physics. But Einstein was to serve a liberating
politics.
    Einstein's fame was consciously exploited to promote the
new stress on modernizations premised on the application of
advanced knowledge. The attractive, heroic image of
Einstein's progressive breakthroughs was used to attract
bright, young people to the sciences and technology.
    Most impressively, however, there no longer was one line
on Einstein. The views put forth seemed almost as varied as
the individuals. One person saw Einstein as a hero who made
the most of skepticism and thus "liberated thought" for all
until his ideas became dogmas which precluded his seeing new
things afresh, a rigidity which supposedly kept him fixated
in his late years on his war with the Copenhagen School.[44]
This view was challenged by someone pointing out that physics
is moving in the direction of Einstein's supposed fixation, a
unified field theory.[45] Someone else criticized Einstein's
philosophy for its metaphysical premises which kept him from
breaking sufficiently with the old physics.[46]
    Zhou Peiyuan defended Einstein's materialistic opposi-
tion to the Copenhagen School, but argued that whatever
Einstein's philosophical flaws might be, they paled before
the bright example of his combination of science and human-

ity, Einstein's refusal to be swayed by narrow materialism; his identification with progressive causes; his opposition to militarism, fascism, Hitler, McCarthyism, and the cold war; and his respect for China's great ancient civilization. Einstein's enemies, Dr. Zhou concluded, were the enemies of all progressive humanity.[47]

For Nanjing University philosopher, Hu Fu-ming, Einstein should simply be comprehended as one of a continuing number of great scientists who proved that all truth is but tentative and open to reexamination, change, progress, and contingent new truths. Einstein by now had been superceded as even the greatest of creators (read Mao) should be.[48]

Zhou Peiyuan, at the Einstein Centennial Conference in Beijing in 1979 held up Einstein's socialism before the Chinese people as the humane democratic way to apply the principles of the Paris Commune and to build socialism. Dr. Zhou cited the conclusion of Einstein's May 1949 Monthly Review essay on why he was a socialist.

> A planned economy is not yet socialism. A planned economy as such may be accompanied by the complete enslavement of the individual. The achievement of socialism requires the solution of some extremely difficult socio-political problems: how is it possible in view of the high degree of concentration of political and economic power, to prevent administrative personnel from becoming all-powerful and overbearing? How can the rights of the individual be protected and therewith a democratic counterweight to administrative power be assured?[49]

Writers in Philosophical Studies on "Einstein and the Copenhagen School" described as a great historical perversion the attempt of the fundamentalists in 1974 to categorize the Einstein-Bohr dispute as one between "idealistic empiricism and a priori idealism." What the debate between the two truly reflected, rather, was the only scientific route to progress in truth, "a contention among a hundred schools of thought." Einstein and Bohr differed fundamentally, but they respected each other greatly. The thirty years of debate illuminated many topics and facilitated scientific progress even if neither party ever settled the basic philosophical question. It is the conditions facilitating the continuation of their struggle which should be focused upon. Truth can only be perfected in a situation where "the hundred schools of thought contend."[50]

I do not know whether this invocation of Einstein on behalf of democratic values and processes reflects a manipulation of his life on behalf of democratic commitments or a deep conviction about that life. In either case, a celebration of his centennial year in which Einstein's life served the humane causes he cared so much about was a most happy feature of the Einstein Centennial Year in China.

Notes

1. Zhou Peiyuan, "Einstein Commemoration Activities," China Reconstructs (July 1979): 21.

2. Ibid., p. 20.
3. Loren Graham, Science and Philosophy in the Soviet Union, (New York: Vintage Books, 1974 [1966, 1971), pp. 69 - 138.
4. J. L. Talman writes,

the French Revolution, the Great War, Versailles, Munich, Auschwitz, Hiroshima, in older times, a great plague, a barbaric invasion, a jacquerie or a religious-social mass explosion, the catastrophic watersheds in those fate-ridden national histories of the Jews, Poles, the Irish — continue to exercise a compulsive history for long periods. They shape patterns of thought and behavior, engrave images and memories, leave behind their new words, proverbs, and similes. Above all, they hypnotize people, making them unable to see what happens to them here and now, forcing them to view the immediate and the actual through the prisms of the traumatic experience. [The Origins of Totalitarian Democracy (New York: Praeger, 1970), pp. 129 -30.]

5. Interview with Yang Chenning, Des Moines, Iowa, December 6, 1979.
6. "For all Chinese nationalists of Mao's generation it was common ground that the development of their consciousness and the weakening of the imperialist grip upon China had been assisted by the First World War." John Gittings in Mao Tse-tung in the Scales of History, ed. Dick Wilson (London: Cambridge University Press, 1977), p. 270.
7. P. A. Y. Gunter, ed., Bergson and the Evolution of Physics (Knoxville: University of Tennessee Press, 1969), pp. 18 - 19.
8. Zhou Peiyuan, "Einstein Commemoration Activities," p. 21, Einstein did not support China's liberation struggle. Yugoslavia's Review ofInternational Affairs (May 1, 1955) reported upon his death (Veljko Ribar, "A Great Scientist"), "When requested for support, he sent his son (a Serb on his mother's side) to comrade Kosta Novakovic, the organizer of the propaganda struggle against the regime of police oppression, with the statement that his name can always be used for this cause." (p. 17).
9. Selected Works of Mao Tse-tung, vol. 1, (Peking: Foreign Languages Press, 1965), p. 118.
10. Selected Works of Mao Tse-tung, vol. 2, (Peking: Foreign Languages Press, 1965), pp. 132 - 83.
11. Mao, vol. 1, p. 300.
12. Sakata Shoichi, Kagakusha to Shakai (Scientists and Society) (Tokyo: Iwanami Shoten, 1972), pp. 347, and 348. I am deeply grateful to Yasuko Kurachi Dower for translating for me this and other Japanese works by Sakata.
13. Salata Shoichi, Supplement of the Progress of Theoretical Physics, (1971), p. 108.
14. "Chairman Mao's Concern for the Affairs of Astronomy," Kuangming ri bao, December 22, 1978, p. 2.
15. Sakata Shoichi, Supplement of the Progress of Theoretical Physics (1971), pp. 110, 125, and 158.
16. Sakata Shoichi, Kagaku to Heiwa no Sozo (Science and the Creation of Peace) (Tokyo: Iwanami Shoten, 1963),

pp. 142 and 333 - 365.
    17.   Sakata Shoichi, <u>Supplement of the Progress of</u>
<u>Theoretical Physics</u>, (1971), p. 169.
    18.   <u>Mao Zedong sixiang wansui</u> (1969), pp. 149 and213.
    19.   <u>Selected Works of Mao Tse-tung</u>, vol. 5, pp. 332 -
333.
    20.   Ibid., p. 382. <u>Cf</u>. p. 383.
    21.   <u>Mao Zedong sixiang wansui</u> (1967), p. 51 <u>Cf</u>. p. 59.
    22.   Ibid., pp. 149 - 50.   <u>Cf</u>. pp. 266, 206 - 09, and
213.
    23.   <u>Mao Zedong sixiang wansui</u> (1969), p. 476.
    24.   Ibid., p. 560.
    25.   <u>Mao Zedong sixiang wansui</u> (1967), pp. 561 - 67.
    26.   Mao (1969), p. 565.
    27.   Ibid., p. 504.
    28.   Sakata Shoichi, <u>Supplement of the Progress of</u>
<u>Theoretical Physics</u> (1971), pp. 185 - 198. In that article
Sakata cited, as he often had earlier, from a June 16, 1867
letter of Engels to Marx. In a paragraph on expected, new
revolutionary upheavals in chemistry, Sakata has Engels
write, "The atom — formerly represented as the limit of
possible division — is now nothing more than a node which
gives rise to qualitative differences when making divisions."
(p. 187). But if the Moscow, Foreign Languages Publishing
House edition of <u>Karl Marx and Frederick Engels: Selected</u>
<u>Correspondence</u> is accurate, Sakata, probably in transcribing,
has gotten Engels wrong. Engels wrote, "The atom — formerly
represented as the limit of divisibility — is now nothing
more than a relation." (p. 227).   For Marx and Engels, the
focus is always on the relational reflexive whole, not the
issue of infinite divisibility.   It was Hegel, Engels wrote,
who discussed an "infinite series of divisions."   "The
molecule as the smallest part of matter capable of <u>indepen-</u>
<u>dent existence</u> is a perfectly rational category, a 'node' as
Hegel put it, in the infinite series of divisions..." That
notion of a molecule makes no sense today. However, Engels
accepted it.   But in contrast to Mao and Sakata, his stress
— and this is the philosophical nub of the matter — was on
the issue of reality as relation not as infinite division.
    29.   <u>Red Flag</u>, 1965, no. 6.
    30.   <u>Red Flag</u>, 1965, no. 9.
    31.   Ibid., pp. 45, and 46.
    32.   Ibid., p. 60.
    33.   His former Japanese colleagues suggested to me that
the politics of the peace movement in Japan limited Sakata in
this period.   It split on pro and anti-Soviet lines among
others. Those associated with Japan's Communist Party opposed China's policy on the Vietnam War. Sakata, wishing to remain effective in Japan, could not escape the pressure not to
identify too closely with the anti-Soviet, pro-Chinese
group. His admiration for Mao and China, however, did not
wane.
    34.   Zhou Peiyuan, "A Milestone in the History of
Science," <u>China Reconstructs</u> (October 1966): 43.
    35.   Zhou Peiyuan's preface to <u>Aiyencitan, wenji</u>, vol.
I, (Peking: Commercial Publishers, 1977), p. 8.
    36.   "Criticize Yao Wenyuan's View of Natural Science,"
<u>Red Flag</u>, 1978, no. 4, p. 66.
    37.   Zhou Peiyuan in <u>Renmin ri bao</u>, November 20,1977,

p. 1.

38.   Ho Zuoxiu and Kuo Hanyong, Kuangming ri bao, January 18, 1978, p. 3.   Kuo Hanying is Kuo Muoruo's son.

39.   Li Kuo, "Criticize Einstein's World View," Natural Dialectics Journal (in Chinese), (1974), 71.

40.   Ibid, p. 72.

41.   "Einstein and the Copenhagen School," Zhexue yanjiu, 1979, no. 5, p. 59.

42.   They did appear in national journals, controlled by Yao Wenyuan.   Cf. "Marxism and Natural Science," Red Flag, 1976, no. 4.

43.   Juo Yu, "It is Impossible to Tamper with Marxism," Red Flag, July 4, 1977, no. 7, translated in CMP-SPRCM-77-72, pp. 69 and 71.

44.   Shen Mingxian, "On the Specific Characteristics of Einstein's Philosophical Thought," Kuangming ri bao, March 8, 1949, p. 4.

45.   Qu Cunrang, "Correctly Evaluate Einstein's Philosophical Thought," Kuangming ri bao, May 17, 1979, p. 4.

46.   Chen Bishu, "Let's Discuss Einstein's Philosophical Thought," Sizhuan University Studies Journal (1979): 26 - 33.

47.   Zhou Peiyuan's speech at the Einstein Centennial Conference, Renmin ri bao, February 21, 1979, p. 3.   And the preface to Einstein's essays in Chinese translation, Aiyencitan, wenji.

48.   Hu Fuming, "On the Mutual Utility of Practice and Theory," Baike zhishi (May 1979):   44 - 49.

49.   Zhou Peiyuan, Renmin ri bao, February 21,1979, p. 3.   In the English original, the italicized terms are "far-reaching centralization," "bureaucracy," and "the power of bureaucracy."

50.   Zhexue yanjiu, 1979, no. 5, pp. 59 - 65.

18.
Einstein's Views on Academic Freedom
KENNETH FOX

> You may bring flowers to my door when the last
> witch-hunter is silenced, but not before.[1]

Albert Einstein sent this ringing statement to the Emergency Civil Liberties Committee the day before his seventy-fifth birthday on March 14, 1954. At that age, Einstein was still strongly and actively concerned with academic freedom as a part of civil liberties. In honor of his birthday, the Committee held a conference on "The Meaning of Academic Freedom" in Princeton on March 13, 1954. Einstein, however, was not present at the conference.

Clark Foreman, director of the Emergency Civil Liberties Committee, had written to Einstein on February 25, 1954. He submitted five questions which he hoped Einstein would "feel free to answer fully," and which would be used "as the basis for the discussions" at the conference. Although Einstein did not attend, he responded to Foreman emphatically, at length, and in detail.

These questions and answers probably comprise the clearest positions on academic freedom expressed by Albert Einstein and have been published in <u>Einstein on Peace</u>.[2] The complete text of one of his responses, however, does not appear in that volume and is given here in full:[3]

> Question: What in your opinion is the best way
> to help the victims of political inquisitions?
> Answer: It is important for the defense of civil
> rights that assistance be given to the victims of
> this defense who in the above-mentioned
> inquisitions have refused to testify, and beyond
> that to all those who through these inquisitions,
> have suffered material loss in any way. In
> particular, it will be necessary to provide legal
> counsel and to find work for them.
> This requires money, the collection and use
> of which should be put into the hands of a small
> organization under the supervision of a person
> known to be trustworthy. This organization
> should be in contact with all groups concerned
> with the preservation of civil rights. In this

way it should be possible to solve this important problem without setting up another expensive fund-raising machinery.[3]

Notes

1. W. L. Laurence, "Einstein Rallies Defense of Rights," The New York Times, March 14, 1954, p. 69 Also quoted in Einstein on Peace, ed. O. Nathan and H. Norden (New York: Schocken Books, 1960), p. 551. There "witch hunt" appears rather than "witch-hunter."

2. This paragraph does not appear in Einstein on Peace.

3. Dr. Otto Nathan, trustee of the estate of Albert Einstein, has read this paper and has written me the following: "There is no doubt that Einstein knew about the American Civil Liberties Union even if he did not mention it in his statement to the Emergency Civil Liberties Committee." I am grateful to him for this insight. I am also grateful to Dr. Nathan for permission to have access to the nonscientific papers and correspondence of Albert Einstein. I appreciate the friendly cooperation and competent assistance afforded me by the staff of the rare books collection in the Princeton University Library, particularly by Ms. M. Pacheco, Ms. J. Preston, and Ms. A. F. Van Arsdale.

Part VIII

Education and Psychology

19.

Einstein and Psychology: The Genetic Epistemology of Relativistic Physics

ROBERT I. REYNOLDS

The interaction between Einstein and psychological theorists of this century has influenced the way in which scientists from both disciplines represent the physical world. Jacob Bronowski contrasts the representations of classical and relativistic physics as a change in the observer's perspective:

> [Newton's] is a God's-eye view of the world: it looks the same to every observer, wherever he is and however he travels. By contrast, Einstein's is a man's-eye view, in which what you see and what I see is relative to each of us. (1973, p. 249).

It is, therefore, not surprising that Einstein posed episte-mological questions to the leading psychologists of his day, and, in turn, used their empirical findings as a basis for discussion of physicalistic concepts. Newton's closest con-fidants were biblical scholars who helped him relate physical relationships to theological interpretation. Einstein, on the other hand, established working relationships with scien-tists from many disciplines, including three of the most important psychologists of this century: Freud, Wertheimer, and Piaget.

In 1932, Einstein wrote to Freud, asking him if he had insight into why man seemed bent on war and self-destruc-tion. Freud replied in a pessimistic tone, basically reiter-ating his principle of Thanatos, or the Death Instinct. As support for the fledgling League of Nations, the Einstein-Freud correspondence was published in 1933 under the title Why War?

Einstein, the physicist Max Born, and Max Wertheimer were close friends while the three were at the University of Berlin. Wertheimer, the founder of Gestalt psychology, was deeply influenced by the discoveries of relativity and quantum theory, as were his students and later expositors Wolfgang Kohler and Kurt Koffka. The Gestaltists observed that a salient feature of our perceptual experience is that objects appear invariant as they move relative to the observer; we do not experience changes in an object's size, shape, color, or brightness despite continual changes of its image at the level of the eye. In his book <u>Natural Philosophy of Cause and Chance</u> (1949), Born makes use of Wertheimer's ideas about size invariance and gestalten, adopting the psychological concept of "observational invariant" in order to describe physicalistic concepts. In letters to Born, Einstein criticized the adoption of the concept of the observational invariant, believing that it did not do justice to relativity theory. I believe that this disagreement was based on Born's inconsistent adoption of Gestalt theory. The latter emphasizes relational determinants of perception in which perceived characteristics of the judged object are determined by characteristics of its immediately surrounding field. The Gestalt organizational principles apply only within local and separate frames of reference, and as such are indeed consistent with the basic tenets of relativity.

Einstein described his theoretical approach as the explication and questioning of unconscious assumptions and hypotheses about the nature of space, time, and the behavior of objects. His concern with the epistemology of such fundamental concepts has provided the basis for an important interaction between spacetime physics and developmental cognitive psychology. In 1928 Einstein presided over the first international course of lectures on philosophy and psychology at Davos, Switzerland. In conversations with the Swiss developmental psychologist Jean Piaget, Einstein suggested that the following epistemological questions concerning the nature of time might be investigated experimentally. Is our intuitive grasp of time primitive or derived? Is it identical with our intuitive grasp of velocity? What bearing do these questions have on the development of the child's conception of time? These questions provided inspiration for Piaget's subsequent investigations. Piaget published the results of these studies in two volumes on the child's conception of time (1946) and motion and speed (1946), and related volumes on the development of space (1948) and geometry (1948). In response to Einstein's questions, it is clear that the concepts of time and space are not a priori categories, but develop from earlier "schemes," or ways of representing the world. More specifically, the concept of time is only gradually derived from the developmentally more primitive concepts of distance and velocity.

Piaget has noted that the ontogenetic development of the child's representation of space and time proceeds in an order just the reverse from the historical development of these concepts' scientific representation. These concepts begin as an undifferentiated spacetime, followed by their progressive differentiation. The earliest stage of temporal development described by Piaget is highly relativistic, with temporal estimation determined by local frames of reference. The earliest stage of spatial development may be represented in terms of topological relations between clusters of environmental stimuli. Space, like time, is neither homogeneous in all directions, nor isotropic in all orientations, nor is it independent of objects being observed. As in the development of the concept of time, children at the earliest stages fail to dissociate space as a structure from its content. The perceptual world of the young child contains clusters of meaningful space derived from the location, number, size, and meaning of objects being observed at any one moment.

During the latter half of his career, Einstein often made use of observations from Piaget's "genetic epistemology" in order to reconcile relativistic phenomena with the "common sense" representation of the physical world. For Einstein, "the whole of science is nothing more than a refinement of everyday thinking. . . . Physical theories try to form pictures of reality and to establish its connection with the wide world of sense impression." While classical physics represents nature in a form consistent with "common sense," relativistic phenomena, and quantum mechanics to an even greater extent, diverge from this goal of scientific inquiry. Einstein approached this problem by considering the development of the concepts of space, time, and objects in the individual. In his book <u>Out of My Later Years</u> (1950), Einstein describes the order in which temporal concepts develop, obviously drawing from Piaget's empirical research. He concluded, along with Piaget, that these concepts do develop such that the representation of everyday experience is different at different stages of one's development. In <u>The World As I See It</u> Einstein discusses the difficulty confronting our adult perspective when we attempt to introspect developmentally more primitive forms of representation;

> We have forgotten what features in the world of experience caused us to form (pre-scientific) concepts, and we have great difficulty representing the world of experience to ourselves without the spectacles of the old, established conceptual interpretation. (1949).

As early as Einstein's first revolutionary paper on special relativity, he distinguished himself by probing the unconscious assumptions underlying classical physics, namely the concepts of absolute space and absolute time. These concepts allow us to locate objects around us independent of their velocities and to arrange events in a unique time sequence. However, the attempt to develop a consistent account of electromagnetic and optical phenomena revealed that observers moving relative to each other with large velocities will coordinate events differently. The inconsistencies confronting twentieth-century physics were resolved by explicating and transcending these assumptions, and embedding the principles of prerelativistic physics within the more encompassing concepts of special and general relativity. From the psychological point of view, such a change in perspective entailed Einstein's freeing himself from the adult vision of the universe, and replacing it with a developmentally earlier and more global representation, one less obvious to adult and scientific eyes.

The historic dialogue between Einstein and exponents of quantum mechanics such as Niels Bohr and Max Born reveals that the latter were profoundly disappointed that Einstein, a cofounder of quantum theory, was not prepared to accept the principles of quantum mechanics as an adequate description of nature. Quantum mechanics provides a description of the physical world, at the atomic level in particular, that is equivalent in many ways to the ontogenetically most primitive representation of space, time, causality, and objects. An analysis of this primitive form of understanding from the standpoint of psychology may, in fact, shed light on the long-standing argument between Einstein and the quantum theorists.

From the perspective of the infant cognizing at the sensori-motor level (birth to two years of age), objects are not endowed with permanence, nor do they exist separate and independent from space, time, and the activity of the observer. To the adult, a moving object appears to move along a continuous trajectory, within a prescribed space, occuring at a particular time, caused by certain preoccuring events, and independent from the act of observation. To the infant at the sensori-motor level, a moving object is seen as a succession of events created anew at each moment. When an object leaves the immediate perceptual awareness of the infant, it ceases to exist; for the infant, out of sight is truly out of mind. The structural features confronting the infant determine whether an object will or will not continue to exist for it. Thus, after it watches a person walk out the door, the infant will look for him where the person was last seen, for example, by his crib. As the infant progresses through the sensori-motor period, he becomes fascinated with the newfound ability to "create" objects and people by the act of turning his attention upon them.

In quantum mechanics, as in sensori-motor awareness, the act of observation creates the mechanical state — the position or momentum — of a particle. Bohr describes this as follows:

> Since the discovery of the quantum of action, we know that the classical ideal cannot be attained in the description of atomic phenomena. In particular, any attempt at an ordering in space-time leads to a break in the causal chain, since such an attempt is bound up with an essential exchange of momentum and energy between the individuals and the measuring rods and clocks used for observation. (1934, 97-8).

For many quantum theorists — those of the Copenhagen School — the change in the observed state of a particle is not an artifact imposed by the limitations of the methods of observation. Victor Guilleman describes the continual creation and annihilation of particles as fundamental to the nature of the physical world: In the representation provided by Feynman diagrams,

> a vacuum is not an empty space, rather it is a seat of continuous activity with virtual particles of many kinds winking in and out of existence. . . . A stationary particle may be thought of as going into, and out of existence while remaining at the same place; one which is moving is annihilated at one point and created anew at the next. (1968, pp. 303, 213).

Since both relativity theory and quantum mechanics describe nature in ways counter intuitive to the adult perspective, we may ask, as did the early quantum theorists, why

Einstein rejected the quantum mechanical representation as a complete description of the physical world. The answer lies in an assumption implicit in both classical and relativistic physics, namely that the object, as an entity separate (if not totally independent) from the space and time in which it exists, is essential to any description of physical reality. In order to establish a connection between physical theories of reality and "the wide world of sense impressions," Einstein once again made use of the observations of genetic epistemology. Piaget points out that the child's construction of reality during the first two years of life culminates in its attributing permanence to objects. In <u>Out of My Later Years</u>, Einstein uses these observations to formulate what was for him the basis of any acceptable description of physical reality:

> I believe that the first step in the setting of a "real external world" is the formation of the concept of bodily objects and of bodily objects of various kinds. Out of the multitude of our sense experiences we take, mentally and arbitrarily, certain repeatedly occuring complexes of sense impressions. . . , and we attribute to them a meaning — the meaning of bodily objects. (1950, p. 59).

For Einstein, the principles of quantum mechanics provide an unacceptable formulation of physical reality since preeminence is given to ψ-functions or wave patterns, rather than to sensible objects. In ontogenetic terms, quantum mechanics provides a representation equivalent to a developmental period prior to the cognizing of a permanent and objective reality. In brief, Einstein's theory of relativity may be described as a child's-eye view of the physical world, while quantum mechanics provides a view characteristic of the infant's preobjective reality.

At this point we may speculate on the act of creation involved in discovering a revolutionary insight. The developmental model that I am proposing is hierarchical in nature, that is, the more primitive levels are not lost, but are embedded within the more advanced. The insights that Einstein attained in the areas of special and general relativity, and in quantum theory, undercut the adult vision of the universe along with its implicit assumptions about the nature of space, time, and objects. The prevalent Newtonian conception was replaced with the ontogenetically more primitive and global representations of relativistic and quantum theories, both of which are embedded within the cognitive structures of all adults.

This analysis leaves unanswered the interesting psychological question of how Einstein, more than any other physicist of his time, came to so many brilliant and revolutionary insights. Why was it Einstein and not Poincare who discovered the relativity of simultaneity? Einstein confided to a friend how he was able to undercut the adult vision of the universe.

> When I asked myself how it happened that I in particular discovered the Relativity Theory, it

seemed to lie in the following circumstance. The
normal adult never bothers his head about
spacetime problems. Everything there is to be
thought about, in his opinion, has already been
done in early childhood. I, on the contrary,
developed so slowly that I only began to wonder
about space and time when I was already grown
up. In consequence I probed deeper into the
problem than an ordinary child would have done.
(Seelig, 1954, p. 71)

I do not believe that it is a mere coincidence that both
Einstein and Newton produced their most creative work while
isolated from academic institutions: Einstein, while working
as a patent clerk, Newton, while at his estate in retreat
from the plague of 1665. In a letter from Einstein to Born
(Born, 1971), Einstein recommended that a promising young
student of Born's find work outside of academia while independently working on theoretical physics. If independence
from institutions propounding the existing paradigm does in
fact accelerate creativity, the question becomes one of
determining the characteristics necessary to sustain self-
directed and self-motivated inquiry. Such personality
characteristics are the study of psychobiographers interested
in the nature of creative genius.

Bibliography

Niels Bohr, Atomic Theory and the Description of Nature
(New York: Macmillan, 1934), p. 9.

Max Born, The Born-Einstein Letters, trans. I. Born, (New
York: Walker, 1971).

Max Born, Natural Philosophy of Cause and Chance,
(Oxford: Clarendon Press, 1949).

J. Bronowski, The Ascent of Man, (London: BBC, 1973).

Albert Einstein, Out of My Later Years (New York:
Philosophical Library, 1950).

Albert Einstein, The World As I See It, trans. Alan
Harris, (New York: Philosophical Library, 1949).

Victor Guilleman, The Story of Quantum Mechanics (New York:
Scribner, 1968).

Jean Piaget et al., Le developpement de la notion de temps
chez l'enfant (Paris: Presses universitaires de France,
1946).

Jean Piaget et al., La geometrie spontance de l'enfant
(Paris: Presses universitaires de France, 1948).

Jean Piaget et al., Les notions de movement et de vitesse
chez l'enfant (Paris: Presses universitaires de France,
1946).

Jean Piaget et al., <u>La representation de l'espace chez
l'enfant</u> (Paris: Presses universitaires de France, 1948).

Carl Seelig, <u>Albert Einstein: eine dokumentarische
Biographie</u> (Zurich: Europa Verlag, 1954).

20.

Epistemological and Psychological Aspects of Conceptual Change: The Case of Learning Special Relativity

GEORGE J. POSNER, KENNETH A. STRIKE,
PETER W. HEWSON, AND WILLIAM A. GERTZOG

A focal point of recent work in the philosophy of science has been the study of conceptual change in science. In our own current work, we were struck with the potential applicability of current philosophical and historical accounts of this process to our research on science learning. A case in point is the conceptual restructuring required of college students learning the special theory of relativity.

We assume that special relativity can be considered a new conception (let us label it c_2) and that learning it is a matter of modifying the student's existing conceptions (let us label them c_1) to accommodate it.

Dissatisfaction with C_1

Before people actively search for alternative conceptions (c_2) they normally must find existing conceptions (c_1) inadequate. This dissatisfaction is not likely to have existed during the period when their c_1 was forming. Rather, dissatisfaction is likely to grow as recognition of c_1's inability to assimilate experience grows.

However, finding usable anomalies for instructional purposes and making them accessible to students is no easy matter. Historical anomalies at first blush are the most logical source to consider. For example, the now famous Michelson-Morley experiment (M-M) presented one such anomaly to prerelativity physicists and is typically presented to students as a prelude to the special theory. However, in order for one to perceive M-M as anomalous, one must understand and accept the experiment in terms of its underlying observational theory (experiments are rarely "transparent"), one must also understand and accept the conception within the context of which the experiment is anomalous (in this case, the ether theory), and, further, one must be able to trace the empirical implications of the conception so as to realize the anamalous nature of the experimental finding.

When faced with an anomaly, the individual (scientist or student) has several alternatives. One may come to the conclusion that his or her existing conceptions require some fundamental revision in order to eliminate the conflict. But this is the most difficult and, therefore, the most unlikely

approach. Other possibilities are rejection of the observation, a lack of concern with experimental findings on the
grounds that they are irrelevant to one's current conception,
a compartmentalization of knowledge to prevent the new information from conflicting with existing belief, or an attempt
to assimilate the new information into existing conceptions.
     The first of these conflict reduction strategies entails
a disbelief in the experimental finding, a strategy not un-
common even among scientists.[1] The student may not interpret
the M-M as a valid measure of the speed of light. With the
second strategy, the student ignores the experiment because
he or she cannot or will not relate it to the conception at
stake. The experiment remains at the periphery of consciousness, on the grounds that there are many unexplainable phenomena in the world, only some of which the student will worry
about. The third strategy involves the construction of categories for information with no requirement that information
in different categories be integrated. The student may
believe that facts about light, electrodynamics, and
mechanics have no necessary relationship to each other. This
is an especially likely strategy if the student does not
share Einstein's commitment to parsimony of physical laws.
The fourth strategy, inappropriate assimilation, may be
illustrated by the student attempting to "Newtonize" all
physical phenomena, for instance, by developing a mechanical
interpretation of the M-M finding similar to the Lorentz contraction theory.

Intelligibility of c_2

     The special theory is generally accessible with respect
to both its terminology (one needs only to comprehend the
terms velocity, coordinate system, uniform translatory
motion, and other such terms known by most beginning physics
students) and its mathematics (the simple algebra necessary
for deriving the transformation equations). Other conceptions are not so intelligible. For instance, Einstein's
general theory of relativity requires knowing the meaning of
Gaussian coordinate systems; a conception of wave-particle
duality requires knowing Schroedinger's wave equations.
     However, as recent research on language comprehension
demonstrates, finding discourse intelligible requires more
than just knowing what the words and symbols mean. Intelligibility also requires constructing or identifying a coherent
representation of what a passage is saying. The representation of a theory functions as a format into which new information must be fit. Information which does not fit is
anomalous. M-M, for example, was an anomaly for the prevailing theories in physics which included a "lumiferous ether."
Sometimes anomalies are "brushed under the carpet" in much
the same way that readers brush anomalous sentences under the
carpet, that is,they don't bother with them. M-M might have
been eventually ignored had it not been for Lorentz's
successful efforts to give an ether theoretical explanation
of the results. (Of course, subsequently Einstein showed
that M-M null results could not have been otherwise according
to special relativity.) Similarly, a representation of the
atom as a mini solar system would not allow a person to make
much sense of quantum changes in the energy of electrons.

Thus, how one represents knowledge and theories determines his ability to make sense of and use the new ideas.
Only if the student can construct a coherent representation
of a theory can it become an object of assessment and a tool
of thought. Only an intelligible theory can be a candidate
for a new conception in a conceptual change.

How difficult is this task for special relativity?
Einstein's two major assertions (his two postulates) can be
stated as follows:

> All universal laws of nature are valid, regard-
> less of an observer's relative motion (so long as
> the motion is uniform translatory motion).

The velocity of light is a constant regardless of the
motion of the observer or the source of the light.[2]
Constructing a coherent representation of each of these
postulates individually is not particularly problematic. One
can imagine a state of affairs in which each in turn is true,
although the more one accepts Newtonian mechanics the harder
it will be to imagine a world in which the second postulate
is true.

The intelligibility of the theory as a whole, however,
is a different story. Finding it intelligible entails
imagining a world in which both of Einstein's postulates are
true, together with the logical implications of the postulates for notions of space and time. This task is a demand-
ing one. The modifications special relativity requires in an
individual's conceptions of space and time are fundamental
— they are almost certain to necessitate a restructuring of
central and strongly held beliefs which in turn would bring
about changes in related belief structures. While far-reaching, some of these changes can also be rather subtle. To
make matters even more difficult, it is possible to apply the
postulates and formulas of special relativity in a superficial way without either having revised one's conceptions of
space and time in accord with the theory or even having
understood the full implications of its principles. Thus,
both learner and instructor can mistake the intelligibility
of the parts — the postulates of the special theory — for the
comprehension of the whole.

In light of such difficulties, special effort is needed
to make relativity fully intelligible to students. One way
in which to make something unfamiliar intelligible is to consider it in the light of something familiar. Thus, metaphors
and analogies have recently gained the attention of philosophers and cognitive psychologists as critical tools of
thought.[3] The attention they have received has been focused
on their function in the process of comprehension as the kind
of coherent representation necessary for making sense of a
set of symbols.

Initial Plausibility of c_2

One source of difficulty in learning relativity stems
from its lack of initial plausibility to physics students.
Initial plausibility can be thought of as the anticipated
degree of fit of a new conception into an existing conceptual
ecology. There appear to be at least five ways by which a

conception can become initially plausible:

1. one finds the new conception capable of solving
    problems of which one is aware,
2. one finds the conception to be consistent with
    other theories or knowledge,
3. one finds the conception to be consistent with
    past experience,
4. one finds or can create images for the conception,
    and
5. one finds it consistent with one's current meta-
    physical beliefs and epistemological commitments.

The last of these is a set of epistemological commitments. Einstein was committed to two fundamental epistemological principles: (1) a theory must not contradict empirical facts; and, (2) the premises of the theory must be characterized by "naturalness" or "logical simplicity," a kind of
"inner perfection" of the theory. He was committed so fully
to these two principles that he was able to apply them ruth-
lessly, even if that application meant a rejection of our
commonsense notions of space and time. Needless to say,
students often do not share Einstein's epistemological
commitments, but their own commitments are likely to be
highly significant in determining what they find initially
plausible and, thus, in shaping their conceptual changes.
Therefore, it is important to find out just what epistemological commitments students have if one wants to understand
what they are likely to find initially plausible and (more
generally) their processes of conceptual change. To the
extent that the students' epistemologies are different than
Einstein's, we should not be surprised at difficulties in
finding the special theory initially plausible and in subsequently accommodating it.

Fruitfulness of c_2

These sorts of fundamental assumptions not only function
to increase or decrease a new conception's initial plausibility, they may also provide reasons for ultimately accommodating or resisting c_2. Just as Einstein's epistemological
commitment to symmetry, simplicity, and empirical verification provided the driving force for his conceptual change (at
the expense of commonsense notions of space and time),
students' idiosyncratic fundamental assumptions may make an
accommodation to c_2 appear persuasive to them. On the other
hand, their fundamental assumptions can operate in an equally
powerful way to block conceptual change.

Another potential driving force for an accommodation, as
powerful as that generated by the individual's fundamental
assumptions, is the resolution of c_1's anomalies by c_2. If
the dissatisfaction with c_1 created by its inability to make
sense of experience is followed by learning of an intelligible and plausible alternative which resolves the anomalies of its predecessor, then a change to this new
conception may be viewed as persuasive.

This will likewise prove difficult if the student is
actively resisting c_2 on epistemological or metaphysical
grounds. As we shall discuss shortly, the firmness of the
student's commitments to fundamental assumptions inconsistent
with c_2 will determine, in large part, how difficult it will

be to convince him of the fruitfulness of c_2, not to mention c_1's inadequacy.

Particularly when faced with strenuous resistance to c_2, accommodations often require more than just a resolution of c_1's difficulties by c_2. As we have shown, c_1's difficulties (e.g., anomalies) may simply not be taken seriously by the student, especially if taking them seriously entails threatening his or her fundamental assumptions. In such cases, and even in cases where the student is not resisting c_2, the apparent fruitfulness of the new conception can function as a persuasive reason for accommodation. Once aware of an intelligible, plausible alternative to c_1 that resolves c_1's anomalies, students may actively attempt to map their new conceptions onto the world; that is, they may attempt to interpret experience with it. If the new conception not only resolves its predecessor's anomalies but also leads to new insights and discoveries, then the new conception will appear fruitful and the accommodations of it will seem persuasive.

It should be clear that for most students the fruitfulness of special relativity will not be apparent. Thus, we cannot claim that viewing the world with a relativistic rather than a Newtonian conception will be of any obvious advantage to students. Students need to be aware of phenomena involving very small particles, very high velocities or very great distances before they can fully realize the fruitfulness of the special theory.

A Model of Accommodation

Our description of the four conditions of a successful accommodation may unintentionally have implied a fairly straight forward linear process beginning in dissatisfaction with Newtonian physics, followed by finding special relativity intelligible, leading to an initial belief in its plausibility, and concluding with the belief that the theory is ultimately fruitful. With this major qualification in mind, let us suggest a model of the accommodation process (figure 1). Let us compare the case of an accommodation by physicists with the case of an accommodation by students learning special relativity. Accommodation by physicists is a composite of two distinct situations: development of theory and theory choice. These may be illustrated by the accommodation of Einstein during his development of relativity and by the accommodation of Einstein's theory by his colleagues.

Figure 1

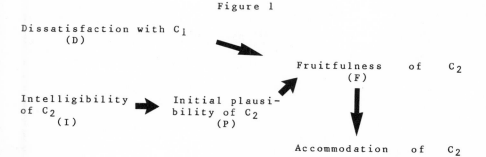

Dissatisfaction with C_1
    (D)

Fruitfulness   of C_2
        (F)

Intelligibility   Initial plausi-
of C_2 bility of C_2
    (I)   (P)

Accommodation   of C_2

In both theory development and theory choice, dissatisfaction with c_1 plays a major role in the process and accounts, to a large extent, for the scientists devoting energy to the development effort or for the scientists taking seriously an alternative to their tradition. On the other hand, intelligibility as a condition is not strictly applicable to theory development but is crucial to theory choice. We can, thus, summarize the theory development or choice case (using the symbols from figure 1) as primarily a $D \longrightarrow$ $I \longrightarrow P \longrightarrow F$ process. Thus, conceptual change in students bears a striking resemblance to conceptual change in science and our understanding of the former will be enhanced through an analysis of the latter.

Strong commitments to scientific conceptions and standards would not be expected in, for example, beginning physics students. How committed can someone be to something he learned only last week? Nevertheless, lurking behind a student's apparent naiveté with regard to physics may lie some fairly firm, but unarticulated, beliefs. A commitment to an absolute frame of reference is so much a part of our culture and is so embedded in our language that it represents a metaphysical belief some twenty years in the making for most college students.[4] There are substantial and significant differences in commitment to conceptions and standards of judgment between physicists and physics students and even among physics students important differences are to be expected.

It follows that while an accommodation for physicists may primarily be a problem of overcoming their prior explicit commitments, for beginning students accommodation may be a problem of finding the new conception intelligible and plausible despite its conflicts with implicit commitments. Second, differences in the strength of students' commitments are likely to explain differences in the difficulty students face when attempting an accommodation. Students with stronger prior commitments will likely require more compelling reasons for an accommodation and will more likely resist the new conception.

With regard to our model in Figure 1, we can conclude that in the learning-of-theories case the process may be symbolized as either an $I \longrightarrow P$ process (if the student has no firm commitments resisting accommodation), or an $I \longrightarrow P \longrightarrow F \longleftarrow D$ process (if prior beliefs that conflict with new conception are resistant to change). Note that in the latter case the student, unlike the scientists, might not need to find C_1 inadequate prior to a consideration of C_2. All that is required is that the student at some point recognizes the advantages of C_2 over C_1 from a comparison of the two conceptions. This might arise from an examination of C_2 and then a contrasting of C_2 with C_1, or it might arise from an analysis of C_1 followed by a presentation of C_2.

Clearly, the critical factor in this process is the firmness and nature of the students' fundamental assumptions. Surprisingly enough, we have been unable to find a body of research about college students (or anybody else, for that matter) in this regard.[5] Interviewing of people studying the special theory of relativity has been chosen to provide evidence for our account of conceptual change. The interviews were conducted with students in a noncalculus,

self-study, self-paced introductory college physics course
who had completed a unit on special relativity, and with
several physics instructors. In an interview, CP (a student)
outlines her belief in absolute time explicitly and repeatedly. In response to a portion of the simultaneity problem
(for which the special theory predicts that two clocks read
different times) she responds:

> (CP)   I mean, how could they change?   Time only
> goes at one rate, right?

After she has read a written explanation showing the derivation of the relativistic prediction, the interviewer
(I) questions her further:

> (I)   And so what about this idea of absolute
> time?
> (CP)   I can't say that's _not_ true.   Yeah, I mean
> absolute time, it just seems time goes on at
> a certain rate everywhere.   It just seems
> natural that it's constant everywhere.

A tutor in the physics course, SL, also shows a firm
Newtonian commitment.[6] In talking about the questions of
shrinking rods and slowing clocks, he says:

> (SL)   I see them as being — as changing their
> length, or changing their time. But I can talk to
> the person who's moving at the same velocity as
> the stick and the clock. He's telling me that
> they don't change. I feel they haven't changed,
> but the way I'm looking at them has changed. I
> guess I'm allowing for the fact that a person
> who's seeing these things at rest, who has his
> clock at rest, his meter stick at rest, has
> [pause] a little more right to say what is really
> happening to the sticks.

A little later he continues:

> (SL)   But I'm not at all uncomfortable with the idea of
> fore-shortening. I do say, I do feel it is a per-
> ception. I will say it is a shortening. I know
> in the back of my mind that my friend who's riding
> along with that meter stick is telling me all the
> time that as far as he can tell, it's the same
> length and I believe what he's saying, which is
> o.k.
> (I)   It's not a conflict?
> (SL)   No, because the fact that it's moving makes it
> appear to me as if it were foreshortened.

Here SL insists on treating length as constant, independent
of frames of reference. He is, thus, led to treat the
special theory's claims concerning the relativity of length
as simply a distortion of perception.
    An example of an epistemological commitment arose in the
context of a general discussion of the problems raised by
relativity, a clear, spontaneous statement of one student's
view of the relationship between theory and experiment. HU

has derived a mistaken result from his view of the relativity principle which implies that pictures taken by two cameras moving past one another at the same instant and of the same two clocks will show different things. The interviewer looks for confirmation of this view:

(I)   So what you're saying is that they wouldn't agree, they couldn't agree, that they'd actually see different things?

(HU)   Right.

(I)   That doesn't bother you?

(HU)   It did at first, but when you think about it and hash it out there really is no reason why we should limit ourselves to one frame of mind. I like to think abstractly and I can see that. I had trouble realizing that lengths would change, too, but you know, I'm game! No, it doesn't bother me. It's just that we don't realize it due to our slow speeds. I tend to agree with with scientific data that's brought up and when they say that an electron — what was that? — a meson, actually goes with the predictions, what can you do? And once you see the facts, you can stretch your imagination.

HU's stated epistemology is simple and empiricist: theories are derived from experimental evidence. Closer examination reveals that he doesn't always operate in this matter. Having seen a result which convinces him that relativity is correct, he is then prepared to apply this theory to derive a result which is patently contradictory and stand by it regardless of his own everyday experience. We have found that people attempt to come to terms with the implausibility of the special theory in ways which fall short of a complete accommodation, thus solving their immediate problems, but creating a hidden problem which might emerge at a later stage.
    Relativism results when the relativity principle is over generalized. From such a standpoint, what anyone sees depends only on their frame of reference and thus there is no necessity for any two people to agree. The relativity principle becomes an immense rug under which anything at all problematic is swept. Leading up to the following excerpt, KL, like HU in the previous example, comes to the same contradictory implication that two cameras will take different pictures simply because they are moving relative to one another:

(I)   You've come up with a result which you have reviewed and you are happy with it, and I am asking you whether [the above implication] is a problem in terms of how you view the world.

(KL)   Well, it certainly is a problem in terms of the way we usually think about the world. I think I want to stick with my results.

(I)   You want to stick with your results but you would like. . . .

(KL)   Yes, it would be nice if it sort of agreed with common sense.

(I)   And common sense says that the cameras should show
        the same thing?

(KL)   But I guess that relativity doesn't always agree
        with common sense.

KL knows that relativity is renowned for results which are
counter intuitive. Even though she seems a little uncomfortable with the contradiction she has derived, she is able to
use her over-generalized relativity principle to gloss over
the crucial distinction between what is contradictory, and
therefore cannot be, and what is merely counterintuitive.

SL provides another example of an attempt to assimilate
the findings of the special theory into an existing conception, in this case in a rather more sophisticated and detailed fashion. As we saw previously, SL reveals a firm commitment to objects with fixed properties and explanations given
in mechanistic terms. He is not alone in this, since Lorentz
in the years preceding 1905 showed a very similar set of
assumptions. What is of interest to us at this point is that
SL reveals this commitment by using it as the conception to
which he assimilates the findings of special relativity. In
order to do this he has to make an auxiliary assumption:
that a shrinking rod constitutes a perceptual problem, and
doesn't actually shrink ("I feel they [rods and clocks]
haven't changed, but the way I'm looking at them has
changed"). This is not necessary or even consistent with an
Einsteinian perspective based on a reanalysis of space and
time. It does, however, play an integral part in protecting
SL's metaphysical commitments.

Assimilation of the findings of special relativity into
existing conceptions serves two purposes. First, it protects
a person's metaphysical beliefs by using conceptions which
are rooted in these beliefs. This generally requires auxiliary assumptions which link findings with beliefs. Second,
it reduces the extent to which any given assertion appears
counter-intuitive.

How a learner decides whether to assimilate or accommodate appears to depend on at least three factors. The
first is the strength and depth of metaphysical beliefs about
the world. The second factor is the nature and strength of
the person's epistemological commitments, for example,
commitments concerning symmetry of physical laws, the
elegance, economy, and parsimony of theory, and the relations
between theory and experience. The third factor is the presence of anomalies with respect to an existing conception.
Such an anomaly might be viewed as an experimental result
which contradicts a prediction, or as an inconsistency or
contradiction which occurs while attempting to assimilate a
result into the existing conception. The extent to which
there is an awareness of both an anomaly and an alternative
conception which resolves the anomaly is a measure of the
extent to which an accommodation is more compelling than an
assimilation.

However, there is little evidence in the interviews
that students were aware of anomalies, even though videotapes
of two experiments which produced results at variance with a
Newtonian viewpoint formed part of the study material in the
course. It is significant that the clearest example of a
student's awareness of anomalous behavior occurs along with a

statement of his epistemological commitment. We find HU saying:

> (HU)   I tend to agree with scientific data that's
> brought up and when they say that an
> electron — what was that? — a meson, actually goes
> with the predictions, what can you do? And once
> you see the facts, you can stretch your
> imagination.

He sees the anomaly, he sees the alternative conception and
his epistemological commitment allows him little alternative
but accommodation. Two features of a conceptual ecology have
the potential for driving the change process from one conception (c_1) to another(c_2): (1) anomalies of c_1, and (2) standards of judgment, based on metaphysical beliefs and epistemological commitments, in terms of which c_1 is seen to contain
inconsistencies, whereas c_2 is not.
   If taken seriously by students, anomalies provide the
sort of cognitive conflict (like a Kuhnian state of "crisis")
that prepares the student's conceptual ecology for an accommodation. The more students consider the anomaly to be
serious, the more dissatisfied they will be with c_1, and the
more likely they will be ready to accommodate c_2, providing
that c_2 successfully explains c_1's anomaly.
   Metaphysical beliefs and epistemological commitments
form the basis on which judgments are made about new knowledge. Thus, a conceptual change will be rational to the
extent that students have at their disposal the requisite
standards of judgment necessary for the change. If a change
to special relativity requires a commitment to the parsimony
and symmetry of physical theories (as it did for Einstein),
then students without these commitments will have no rational
basis for such a change. Faced with such a situation,
students if they are to accept the theory, will be forced to
do so on nonrational bases, for example, because the book or
the professor says it is "true."
   The history of science reveals that many conceptual
changes in science have been driven by the scientists' fundamental assumptions rather than by the awareness of empirical
anomalies.[7] Einstein's special relativity can be seen as
such a case.[8] However, since it is unlikely that students in
an introductory physics course can be successfully taught the
requisite standards of judgment for an accommodation of
special relativity, physics teachers must rely on these
"anomalies" (in the retrospective sense) to drive the accommodation.
   But most of the anomalies will not be readily seen as
anomalies by students without a thorough understanding of the
observational theory in which the experiment was embedded.
Does this problem mean that the special theory can realistically be made at best only intelligible and possibly plausible, but never persuasive to students who are firmly committed to a set of conflicting metaphysical beliefs and epistemological commitments? Is it realistic to expect science
instruction to produce accommodation in students rather than
merely to help students make sense of new theories? And
secondarily, should this be so for all students, or only for
certain groups, such as science majors?

We believe that science teachers should aim at developing in students:

1. an awareness of their fundamental assumptions and of those implicit in scientific theory,
2. a demand for consistency among their beliefs about the world, and
3. an awareness of the epistemological and historical foundations of modern science.

The course content should be such that it renders scientific theory intelligible, plausible, fruitful, and ultimately persuasive. More emphasis should be given to assimilation and accommodation by students of that content than to content "coverage." "Retrospective anomalies" should be included, particularly if historically valid anomalies are difficult to comprehend, or, as with the special theory, were not responsible for driving the conceptual change.[9] Sufficient observational theory should be taught for students to understand the anomalies employed. Any available metaphors, models, and analogies should be used to make a new conception more intelligible and plausible.

Teaching is typically thought of as clarifying content presented in texts, explaining solutions to problems, demonstrating principles, providing laboratory exercises, and testing for recall of facts and ability to apply knowledge to problems. That is, teaching is for recall and assimilation. For teaching aimed at accommodation the following possible changes in this approach are implied:

1. Develop lectures, demonstrations, problems, and labs that can be used to create cognitive conflicts in students. Among other things, one might consider what types of homework problems would create the kind of cognitive conflict necessary as preparation for an accommodation, and whether labs could be used to help students experience anomalies.
2. Organize instruction so that teachers can spend a substantial portion of their time in diagnosing errors in student thinking and identifying defensive moves used by students to resist accommodation.
3. Develop the kinds of strategies that teachers could include in their repertoire to deal with student errors and moves that interfere with accommodation.
4. Help students make sense of science content by representing content in multiple modes (e.g., verbal, mathematical, concrete practical, pictorial),[10] and by helping students translate from one mode of representation to another.
5. Develop evaluation techniques to help the teacher track the process of conceptual change in students (e.g., the Piagetian clinical interview).

The teacher as clarifier of ideas and presenter of information is clearly not adequate for helping students accommodate new conceptions. Our research suggests that the teacher might have to assume two further roles in order to facilitate student accommodation. In these roles the teacher would become:

1. An adversary in the sense of a Socratic tutor. In
    this role, the teacher confronts the students with
    the problems arising from their attempts to assimi-
    late new conceptions. A point of concern is the
    need to avoid establishing an adversarial role with
    regard to students as persons while developing and
    maintaining it with regard to conceptions.
2. A model of scientific thinking. Aspects of such a
    model might include a ruthless demand for consis-
    tency among beliefs and between theory and empirical
    evidence, a pursuit of parsimony among beliefs, a
    skepticism for excessive "ad hocness" in theories
    and a critical appreciation of whether discrepancies
    between results and theory constitute an anomaly or
    are insignificant enough that results may be in
    "reasonable agreement" with theory.[11]

Notes

1.   Compare Lakatos' account of the Michelson-Morley
experiment in "Scientific Research Programmes," pp. 159 -
165, and Holton's account of the Millikan experiment in
"Electrons or Subelectrons? Millikan, Ehrenhaft and the Role
of Preconceptions," in History of Twentieth Century Physics,
C. Weiner, ed., New York: Academic Press, 1977, pp. 266-89.
2.   This statement of the postulates of special relativity is a paraphrase of Einstein's presentation in Albert
Einstein, "What is the Theory of Relativity?" in Ideas and
Opinions, trans. and rev. Sonja Bargmann (New York: Crown
Publishing Co., 1954; reprint ed., New York: Dell Publishing
Co., 1973), pp. 224 - 25.
3.   See Marc Belth, The Process of Thinking (New York:
David McKay, 1977) and Andrew Ortony, "Why Metaphors Are
Necessary and Not Just Nice," Educational Theory 25, (1975):
45 - 53.
4.   See Benjamin Lee Whorf, Language, Thought, and
Reality, ed. John B. Carroll (Cambridge, Mass.: MIT Press,
1956).
5.   An exception, however, is William G. Perry, Jr.,
Forms of Intellectual and Ethical Development in the College
Years: A Scheme (New York: Holt, Rinehart, and Winston,
1970).
6.   It is generally assumed by all textbook writers that
students would start their study of special relativity from a
Newtonian perspective. However, there is evidence that in
many cases this is not so, and that many students remain convinced Aristotelians. (John Lochhead, personal communication.)
7.   See Burtt's account of Copernicus, whose theory was
not a response to anomalies, but was only presented as a
simpler and more harmonious interpretation. Edwin Arthur
Burtt, The Metaphysical F-undations of Modern Science, rev.
ed. (Garden City, N.J.: Doubleday and Co., Doubleday Anchor
Books, 1932).
8.   Empirical findings anomalous with respect to Newtonian physics but consistent with Einsteinian theory developed
many years after the special theory of relativity was proposed.
9.   See Anthony P. French, Special Relativity (New York:

W. W. Norton Co., 1968), pp. 6 - 29, for an example of the use of retrospective anomalies in teaching special relativity.

   10.   See John J. Clement, "Some Types of Knowledge Used in Understanding Physics," University of Massachusetts, Department of Physics and Astronomy, 1977. (Mimeographed.)

   11.   See Thomas S. Kuhn, "The Function of Measurement in Modern Physical Science," in T. Kuhn, <u>The Essential Tension</u>, Chicago: University of Chicago Press, 1977, pp. 178 - 224, for an insightful discussion of this point.

Albert Einstein As An Intercultural And Interdisciplinary Phenomenon: His Influence In All Fields Of Thought

International Conference Celebrating the 100th Anniversary of the Birth of Albert Einstein.

1879 - 1979

NOVEMBER 8, 9, 10, 1979

HOFSTRA
UNIVERSITY
HEMPSTEAD, NEW YORK 11550

CONFERENCES AT HOFSTRA UNIVERSITY

Publication Dates:

George Sand Centennial – November 1976   Vol. I – Fall 1979

Heinrich von Kleist Bicentennial – November 1977   Vols. III, IV – Spring 1980

The Chinese Woman – December 1977

George Sand: Her Life, Her Works, Her Influence –   Vol. II – Spring 1980
    April 1978

William Cullen Bryant and His America – October 1978   Vol V – 1980

The Trotsky-Stalin Conflict and Russia in the 1920's   1980
    March 1979

Albert Einstein Centennial – November 1979   Vol. VI – 1981

Sean O'Casey – March 1980   Spring 1981

Walt Whitman – April 1980   Vol. VII – 1981

Nineteenth Century Women Writers – November 1980   Vols. VIII & IX – 1982

Fedor Dostoevski – April 1981   Vol. X – 1983

Gotthold Ephraim Lessing – November 1981

Johann Wolfgang Goethe – 1982

Ivan Turgenev – 1982

Karl Marx – 1983

George Orwell – 1984

Sponsored by the University Center for Cultural & Intercultural Studies (UCCIS)

Einstein as drawn by the Russian painter, Leonid Pasternak.
    Reproduced by permission of Kinder Verlag, Munich.
From Einstein and the Generations of Science, Lewis S. Feuer
    (Basic Books, Inc., New York, NY).

PROGRAM

Conference Committee:

Joseph G. Astman

Frank S. Lambasa

John T. Marcus

James A. Moore

Conference Coordinators:

Natalie Datlof

Alexej Ugrinsky

Cooperating Agencies:

Institute for Advanced Study, Princeton, NJ
    Harry Woolf, Director
    Mary Wisnovsky, Coordinator, National Einstein Celebration

Nassau County Office of Cultural Development, Roslyn, NY.
    Marcia E. O'Brien, Director

Nassau Library System, Uniondale, NY
    Andrew Geddes, Director
    Gloria Glaser, Director of Public Relations

Thursday, November 8, 1979

1:00 - 3:00 P.M.   Registration David Filderman Gallery
                                        Dept. of Special Collections
                                        Hofstra University Library
                                        9th Floor

3:00 - 5:00   Opening Remarks and Greetings:

                              Joseph G. Astman, Director
                              University Center for Cultural &
                              Intercultural Studies

                              Walter Fillin, President
                              Hofstra Library Associates

                              David Rothman, Southold, NY

                              Special Address: "Albert Einstein as a
                                                 Human Being."

                          Reception

                          Opening of the Albert Einstein Exhibit

                          Dinner   Student Center Cafeteria
                                   $3.85 prix fixe

Thursday, November 8, 1979, cont'd.

7:00 P.M.   Evening Program — Dining Rooms ABC
                                              Student Center
                                              North Campus

                             Albert Einstein in Film

                             Introduction: Mary Wisnovsky, Coordinator
                                            National Einstein Celebration
                                            The Institute for Advanced Study
                                            Princeton, NJ

                             "Albert Einstein: Education of a Genius."
                                Part of the National Einstein Traveling
                             Exhibit sponsored by the Institute for Advanced
                             Study in Princeton, NJ and prepared by the
                             American Institute of Physics, New York, NY

                             "Working with Einstein."
                                The taped conference proceedings at the
                             Institute for Advanced Study in Princeton, NJ

                             Dramatic Presentation: First Performance

                             Introduction: Natalie Datlof
                                            Conference Coordinator
                                            University Center for Cultural &
                                            Intercultural Studies

                             ┌───┐
                             │ Ed Metzger as   │
                             │ ALBERT EINSTEIN: THE PRACTICAL BOHEMIAN   │
                             │   An original one-man play   │
                             │   │
                             │ Produced and Directed by Laya Gelff   │
                             │   │
                             │ Written by Ed Metzger and Laya Gelff   │
                             │   │
                             │ Originally staged at the Players Theatre,   │
                             │ New York, NY and the Matrix Theatre,   │
                             │ Los Angeles, CA   │
                             └───┘

                             Sponsors:

                             Hofstra College of Liberal Arts & Sciences
                             Robert C. Vogt, Dean

                             Nassau County Office of Cultural Development
                             Marcia E. O'Brien, Director

Friday, November 9, 1979

8:00 – 9:00 A.M.   Registration   Dining Rooms ABC
                                                   Student Center

                                   Coffee

9:00 – 9:30   Greetings from the Hofstra University Community

                                   Robert C. Vogt, Dean
                                   Hofstra College of Liberal Arts and Sciences

9:30 – 10:30   Stanley L. Jaki, Seton Hall University
                                                  S. Orange, NJ

                                   Opening Address: "Einstein's Theories: Or the
                                                    Absolute Beneath the Relative."

10:30 – 12:00   PANEL I: HISTORY AND PHILOSOPHY OF SCIENCE

                                   Moderator: Esther B. Sparberg
                                               Department of Chemistry
                                               Hofstra University

                                   "Relativity before Einstein: Leo Hebraeus and
                                   Gianbattista Vico."
                                   William Melczer, Syracuse University
                                   Syracuse, NY

                                   "Einstein, Extensionality and the Principle
                                   of Relativity."
                                   Patrick J. Hurley, University of San Diego,
                                   San Diego, CA
                                   Dennis A. Rohatyn, University of San Diego,
                                   San Diego, CA

                                   "Einstein and Anthropocentricity and Solipsism
                                   in Scientific Philosophy."
                                   Joseph LaLumia, Hofstra University,
                                   Hempstead, NY

12:00 – 1:00   Lunch   Student Center Cafeteria
                                           Dining Rooms ABC

Friday, November 9, 1979, cont'd.

1:00 - 1:30 P.M.
      Introduction: James A. Moore
                      Dept. of Physics
                      Hofstra University

      Heinz Pagels, Rockefeller University,
                New York, NY
                Einstein Lecture Bureau of the
                Institute for Advanced Study
                Princeton, NJ

      Special Address: "Contemporary Physics and
                    Einstein's Vision of
                    Unification."

1:30: - 3:00 P.M.
      PANEL II: METAPHYSICS

      Moderator: William P. McEwen
                Distinguished Service Professor
                Emeritus of Philosophy
                Hofstra University

      "The Nature of Causality."
      Richard Dobrin, Sarah Lawrence College
      Bronxville, NY

      "Einstein and the Limits of Reason."
      Richard Fleming, University of Kansas,
      Lawrence, KS

      "A Comparative Interpretation of Einstein's
      Metaphysics."
      Ashok K. Gangadean, Haverford College
      Haverford, PA

      "The Situation of the Young Einstein -- The
      Influence on His Thinking."
      Wilhelm K. Essler
      Johann Wolfgang Goethe University
      Frankfurt, Federal Republic of Germany

3:00 - 3:30
      Coffee

Friday, November 9, 1979, cont'd.

3:30 - 5:30   PANEL III: LITERATURE

                             Moderator: Rhoda Nathan
                                          Department of English
                                          Hofstra University

                             "Springtime of the Mind: Poetic Responses to
                             Einstein and Relativity."
                             Carol Donley, Hiram College, Hiram, OH

                             "The Circuitous Path: Einstein and the
                             Epistemology of Fiction."
                             Robert Hauptman, University of Pittsburgh,
                             Pittsburgh, PA
                             Irving Hauptman, SUNY at F.I.T., New York, NY

                             "A Search for Form: Einstein and the Poetry
                             of Louis Zukofsky and William Carlos Williams."
                             Stephen R. Mandell, Drexel University,
                             Philadelphia, PA

                             "Relativity Theory and T.S. Eliot's 'Four Quartets.'"
                             Jonel C. Salle, University of Kentucky,
                             Lexington, KY

6:30   Cash Bar   University Club

8:00   Albert Einstein Memorial Banquet

                             Introductions and Greetings:

                             Harold E. Yuker, Provost
                             Hofstra University

                             R. Buckminster Fuller, Philadelphia, PA

                             Keynote Address: "The Cosmological Revolution
                                                Brought by Albert Einstein."

Saturday, November 10, 1979

8:00 - 9:00 A.M.   Breakfast Dining Rooms ABC, Student Center

9:00 - 10:30   PANEL IV: RELIGION

                         Moderator: John T. Marcus
                                    Department of History
                                    Hofstra University

                         "Religion, Relativity and Common Sense."
                         William F. Lawhead, Trinity College,
                         Deerfield, IL

                         "Albert Einstein: The Methodological Unity
                         Underlying Science and Religion."
                         Roy D. Morrison II, Wesley Theological Seminary
                         Washington, D.C.

                         "Einstein and African Religion and Philosophy:
                         The Hermetic Parallel."
                         Charles A. Frye, Hampshire College, Amherst, MA

10:30 - 11:00   Coffee

11:00 - 11:30   Introduction: William A. McBrien
                                        Department of English
                                        Hofstra University

                         Dorothy Michelson Livingston, New York, NY

                         Special Address: "Michelson and Einstein:
                                            Artists in Science."

11:30 - 1:00   PANEL V: ETHICS AND EPISTEMOLOGY

                         Moderator: Evelyn Shirk
                                     Department of Philosophy
                                     Hofstra University

                         "Some Ideas about Einstein's Background-
                         Problem."
                         Sheldon Richmond, Acadia University
                         Wolfville, Nova Scotia, Canada

                         "The Moral Implications of Relativity."
                         Keith R. Burich, Canisius College
                         Buffalo, NY

                         "Einstein and Epistemology."
                         Burton H. Voorhees, University of Alberta
                         Edmonton, Alberta, Canada
                         Joseph R. Royce, University of Alberta
                         Edmonton, Alberta, Canada

Saturday, November 10, 1979, cont'd.

1:00 - 2:00 P.M.   Lunch Student Center Cafeteria
                            Dining Rooms ABC

2:00 - 3:30   PANEL IV: POLITICS

                           Moderator: Paul F. Harper, Chairman
                                       Department of Political Science
                                       Hofstra University

                           "Einstein on War and Peace."
                           Thomas Renna, Saginaw Valley State College,
                           University Center, MI

                           "Einstein and Mao: Why Maoism is Perceived as
                           Irrational."
                           Edward Friedman, University of Wisconsin-Madison
                           Madison, WI

                           "Einstein's Views on Academic Freedom."
                           Kenneth Fox, University of Tennessee,
                           Knoxville, TN

3:30 - 4:00   Coffee

4:00 - 5:30   PANEL VII: EDUCATION AND PSYCHOLOGY

                           Moderator: Jerome Notkin
                                       Department of Special Studies
                                       New College, Hofstra University

                           "Einstein and Psychology: The Genetic
                           Epistemology of Relativistic Physics."
                           Robert I. Reynolds, formerly at
                           Pahlavi University, Shiraz, Iran

                           "Learning Special Relativity: A Study of
                           Intellectual Problems Faced by College Students."
                           William Gertzog, Cornell University, Ithaca, NY
                           Peter W. Hewson, Cornell University, Ithaca, NY
                           George J. Posner, Cornell University, Ithaca, NY
                           Kenneth A. Strike, Cornell University, Ithaca, NY

5:30 - 7:00   Dinner Student Center Cafeteria
                            Rathskeller

Saturday, November 10, 1979, cont'd.
_____

7:00 P.M.   A Dramatic Presentation: Second performance
                             _____ Dining Rooms ABC
                                                       Student Center

                             Introduction: Alexej Ugrinsky
                             _____   Conference Coordinator
                                            University Center for Cultural
                                            & Intercultural Studies

┌──┐
│ Ed Metzger as   │
│   │
│ ALBERT EINSTEIN: THE PRACTICAL BOHEMIAN   │
│   │
│   An original one-man play   │
│   │
│ Produced and Directed by Laya Gelff   │
│   │
│ Written by Ed Metzger and Laya Gelff   │
│   │
│ Originally staged at the Players Theater,   │
│ New York, NY and the Matrix Theatre,   │
│ Los Angeles, CA   │
└──┘

                             Sponsors:

                             Hofstra College of Liberal Arts & Sciences
                             Robert C. Vogt, Dean

                             Nassau County Office of Cultural Development
                             Marcia E. O'Brien, Director

9:00 - 11:00   Wine & Cheese Party
                             _____

                             Closing remarks and final gathering

_____

Conference Lounge
_____

Friday, November 9 - Room 142, Student Center - 9:00 A.M. - 4:00 P.M.

Saturday, November 10 - Room 142, Student Center - 1:00 P.M. - 7:00 P.M.

CONFERENCE PARTICIPANTS

Joseph G. Astman	Department of Comparative Literature and Languages, Director University Center for Cultural and Intercultural Studies, Hofstra University
Keith R. Burich	Office of Executive Vice President, Canisius College, Buffalo, NY
Natalie Datlof	Assistant to the Director, University Center for Cultural & Intercultural Studies, Hofstra University
Richard Dobrin	Department of Physics, Sarah Lawrence College, Bronxville, NY
Carol Donley	Department of English, Hiram College, Hiram, OH
Wilhelm K. Essler	Department of Philosophy, Johann Wolfgang Goethe University, Frankfurt, Federal Republic of Germany
Walter Fillin	President, Hofstra Library Associates, Hofstra University
Richard Fleming	Department of Philosophy, University of Kansas, Lawrence, KS
Kenneth Fox	Dept. of Physics & Astronomy, Univ. of Tennessee, Knoxville, TN
Edward Friedman	Dept. of Political Science, Univ. of Wisconsin, Madison, WI
Charles A. Frye	Enfield House, Hampshire College, Amherst, MA
R. Buckminster Fuller	Philadelphia, PA
Ashok K. Gangadean	Department of Philosophy, Haverford College, Haverford, PA
Laya Gelff	Sherman Oaks, CA
William Gertzog	Department of Education, Cornell University, Ithaca, NY
Paul F. Harper	Chairman, Department of Political Science, Hofstra University
Irving Hauptman	Dept. of Physics & Physical Science, SUNY at F.I.T., N.Y., N.Y.
Robert Hauptman	Dept. of Comp. Literature, Univ. of Pittsburgh, Pittsburgh, PA
Peter W. Hewson	Department of Education, Cornell University, Ithaca, NY
Patrick J. Hurley	Department of Philosophy, University of San Diego, San Diego, CA
Stanley L. Jaki	Seton Hall University, South Orange, NJ
Joseph LaLumia	Department of Philosophy, Hofstra University
Frank S. Lambasa	Chairman, Dept. of Comp. Literature & Languages, Hofstra Univ.
William F. Lawhead	Department of Philosophy, Trinity College, Deerfield, IL
Dorothy Michelson Livingston	New York, NY
Stephen R. Mandell	Dept. of Humanities-Communications, Drexel Univ., Philadelphia, PA

John T. Marcus	Department of History, Hofstra University
William A. McBrien	Department of English, Hofstra University
William P. McEwen	Professor Emeritus of Philosophy, Hofstra University
William Melczer	Dept. of Foreign Language & Literature, Syracuse Univ., Syracuse, NY
Ed Metzger	Sherman Oaks, CA
James A. Moore	Department of Physics, Hofstra University
Roy D. Morrison II	Wesley Theological Seminary, Washington, DC
Rhoda Nathan	Department of English, Hofstra University
Jerome Notkin	Department of Special Studies, New College, Hofstra University
Heinz Pagels	Department of Physics, Rockefeller University, New York, NY
George J. Posner	Department of Education, Cornell University, Ithaca, NY
Thomas Renna	Chairman, Department of History, Saginaw Valley State College, University Center, MI
Robert I. Reynolds	Institute for Cognitive Studies, Rutgers University, Newark, NJ
Sheldon Richmond	Department of Philosophy, Acadia University, Wolfville, Nova Scotia, Canada
Dennis A. Rohatyn	Department of Philosophy, University of San Diego, San Diego, CA
David Rothman	Southold, NY
Joseph R. Royce	Center for Advanced Study in Theoretical Psychology, University of Alberta, Edmonton, Alberta, Canada
Jonel C. Sallee	University Honors Program, University of Kentucky, Lexington, KY
Evelyn Shirk	Department of Philosophy, Hofstra University
Esther B. Sparberg	Department of Chemistry, Hofstra University
Kenneth A. Strike	Department of Education, Cornell University, Ithaca, NY
Alexej Ugrinsky	Assistant to the Director, University Center for Cultural & Intercultural Studies, Hofstra University
Robert C. Vogt	Hofstra College of Liberal Arts & Sciences, Hofstra University
Burton H. Voorhees	Center for Advanced Study in Theoretical Psychology, Univ. of Alberta, Edmonton, Alberta, Canada
Mary Wisnovsky	Coordinator, National Einstein Celebration, The Institute for Advanced Study, Princeton, NJ
Harold E. Yuker	Provost, Hofstra University

GREETINGS

I congratulate Hofstra University on its good sense of celebrating the centennial of Einstein's birth in this Einstein Centennial Year. One would think that any university worthy of that name would do so.

But for me, who had the good fortune of visiting Einstein on more than a score of occasions, Albert Einstein was even more than perhaps the greatest scientist of the modern world and the revolutionizer of modern science. He was one of the most superb human beings it has ever been my privilege to meet. Goose-pimples used to go up and down my spine every time I came into his presence. Yet he was one of the humblest, most unassuming of persons.

Do not ask me to explain this fact, for, I can not. It was simply the way he always affected me.

His passing was an inexplicably great loss to all mankind. We shall not live to see his like again.

> Paul Arthur Schlipp, Editor
> The Library of Living Philosophers
> Southern Illinois University at Carbondale
> Carbondale, IL

No doubt that this is a significant meeting. I am so sorry that I am not able to attend. Wishing you great success.

> Yu Kuang Yuan, Deputy Dean
> The Chinese Academy of Social Sciences
> Beijing, China

With best wishes for your conference.

> Chou Pei-yuan, Vice President
> Academia Sinica of the People's Republic
> of China
> Beijing, China

With best regards and best wishes for the upcoming Exhibit and Conference.

> Edward Yen, President
> The Physical Society of the Republic of China
> Professor, Department of Physics
> National Tsing Hua University
> Hsinchu, Taiwan,
> Republic of China

...wishing you a successful International Einstein Conference.

> Klaus-J. Brederhoff
> Vieweg
> Friedrich Vieweg & Sohn
> Verlagsgesellschaft
> Wiesbaden, Federal Republic of Germany

I hope that it will be a very successful occasion.

> A. P. French
> Department of Physics
> Massachusetts Institute of Technology
> Cambridge, MA

I wish you all success with the meeting and will be looking forward with great interest to reading the proceedings.

> Robert Jastrow, Director
> National Aeronautics & Space Administration
> Goddard Space Flight Center
> New York, NY

Please accept my best wishes for a very successful conference.

> Tony Auth
> The Philadelphia Inquirer
> Philadelphia, PA

With best wishes for a successful and exciting conference.

> Oscar Berger
> New York, NY

THE CONSUL GENERAL
OF THE FEDERAL REPUBLIC OF GERMANY
Hartmut Schulze-Boysen

460 Park Avenue
New York, NY 10022

October 19, 1979

Albert Einstein, whose 100th birthday we are commemorating this year, never evaded the social, ethical and political issues of his time. In the course of his life he was deeply involved in the political upheavals and catastrophes of this century. Einstein was forced to leave Germany and found work and a new home in the United States. His thoughts and actions show that he perceived the specific social responsibility of the scientist.

The conference of Hofstra University, which tries to explore Albert Einstein's influence in all fields of thought, promises to be an outstanding cultural event. On this occasion, I convey to you my very best wishes.

CREDIT for the success of the Conference goes to more people than can be named on this program, but those below deserve a special vote of thanks:

HOFSTRA UNIVERSITY OFFICERS: James M. Shuart, President
                               Peter D'Albert, Special Assistant to the President
                               Harold E. Yuker, Provost
                               Robert C. Vogt, Dean, HCLAS

UCCIS: Marilyn Seidman, Conference Secretary
         Sharon Smith, Student Assistant

OFFICE OF THE SECRETARY: Robert D. Noble, Secretary
                         Armand Troncone
                         Doris Brown and Staff

HOFSTRA UNIVERSITY LIBRARY: Charles R. Andrews, Dean

DAVID FILDERMAN GALLERY: Department of Special Collections
                         Marguerite M. Regan, Director
                         Nancy Herb
                         Anne Rubino

HOFSTRA LIBRARY ASSOCIATES: Walter Fillin, President

UNIVERSITY RELATIONS: Harold Klein, Director
                      Brian Ballweg, Media Assistant

SCHEDULING OFFICE: Margaret Shields

SPECIAL SECRETARIAL SERVICES: Stella Sinicki, Supervisor

UNIVERSITY CLUB: Cy Settleman, Manager

ARA SLATER

Index

absolutism 5, 87
absurdists 125
academic freedom 165
acceptance 51
accommodation 181
Addison, Joseph 11
African religion 59
agnosticism 47
anatomism 93
Aristotelianism 99
atomism 93
awe 85
beauty 27, 101
Beckett, S. 129
Bergson, Henri 152
Berkeleianism 20
Biao, Lin 151
bipolarity 90
Bohm, David 79
Bohr, Niels 73, 121, 171
Born, Max 171
Bridgman, P. W. 109, 130
Buddhism 90
Bunge, M. 30
Camus, A. 128
Capra, Fritjof 59
Carnap, Rudolf 6
Cartesianism 99, 101
categories 60, 87
Catholicism 39
causality 100, 171
cause and effect 64
change 156
Charlier, C. V. L. 12
Cheseaux, Jean 11
Ci, Xun 152
civil liberties 165
common sense 110
conceptual change 177
Condorcet 8
consciousness 94
Coover, Robert 131
Copenhagen School 172
cosmic illusion 76
cosmic religion 21, 49
cosmology 11, 75, 77
creativity 27, 174
Credo 47
cubism 137
Cultural Revolution 156
Cummings, E. E. 121
Darwin, C. 8
democracy 146
determinism 107
dialectics 156
Dialoghi d'amore 100
Dickens, C. 7
Dirac, P. A. M. 27

dualism 75, 91, 79
Durrell, Lawrence 130
education 143, 146
Eliot, T. S. 126
empiricism 81, 108
Enlai, Zhou 159
Epistemology 25
EPR experiment 74
ether 2
ethics 8, 22, 83, 121
evolution 7
extensionalism 108
Feigl, Herbert 6
Feuer, Lewis 22
field theory 91
Fitzgerald, Frances 7
FitzGerald, George 8
FitzGerald-Lorentz
    contraction 1
Frank, Philipp 5, 26, 131
free verse 123, 135
Freud, S. 169
Frost, Robert 120
Gang of Four 151
Gedankenexperiments 29
gender 64
general relativity 10
genetic epistemology 171
Gestalt psychology 169
Godel, Kurt 13
Goldstein, Rabbi 39
Halley, Edmund 11
harmony 110, 125
Hartsoeker, Nicolaas 11
Hebraeus, Leo 99
Heisenberg, W. 54, 73,
    90, 153
Hermetic tradition 66
hidden parameters 21
Hinduism 88
history 102
Hodge, Charles 37
Hodson, Geoffrey 59
Hoffmann, Banesh 8, 29
Holton, Gerald 6
hope 120
Howard, Henry 39
human nature 144
human values 145
Hume, David 50, 83
Huxley, T. H. 8
imagination 37
indeterminacy 73
intuition 25
invariance 108
Jahn, Janheinz 60
Jaina 78
janusian thinking 63

Judaism 39
Kafka, Franz 125
Kant, I. 12, 50
Kelvin, W. 11
Krishnamurti 79
Kuhn, Thomas 27
Lambert, Johann 11
Lashchyk, Eugene 112
leaps forward 154
Lem, Stanislaw 131
Lenin, V. I. 19
life force 60
logical positivism 6, 81
loneliness 120
Lorentz, H. A. 1, 8
Lorentz-Fitzgerald contract-
  ion 109
love 100
Mach, Ernst 1, 19
MacLeish, Archibald 120
macrocosm and microcosm 62
Maoism 151, 153
Margenau, Henry 27, 52
Marxism 156
Maupertuis 8
Maxwell, James Clerk 5
meaninglessness 129
Michelson, A. A. 1, 6
Michelson-Morley exp 9, 177
mind 60, 120
monism 107
Morley, Edward 1
Mozart 2
mystery 121, 135
Nagarjuna 90, 94
Neo-Kantians 26
Newton, I. 9, 135
Newtonianism 183
non-finito universe 99
Northrop, F. S. C. 51
O'Connell, Cardinal 38
objectivism 138
objectivity 53
observation 172
Olbers, Heinrich 11
Olson, Charles 122
ontology 77, 82, 87, 173
pacifism 143
paradigm crossing 92
Peiyuan, Zhou 151
perceptions 108
phenomenalism 112

politics 143
positivism 5, 23, 112
Pound, Ezra 121
prophet 147
Pynchon, Thomas 130
rationality 81
realism 6, 37
reality 106, 110
Reichenbach, Hans 6, 26
relativism 89, 184
relativization 10
religion 21
Renaissance 99
Russell, Bertrand 42, 151
Rutherford, Ernest 119
Sakata Shoichi 153
Sartre, J. P. 128
Schopenhauer, Arthur 51
Seeliger, Hugo 11
Shah, Idries 32
Shanghai Gang 157
simultaneity 137, 173
Solovine, Maurice 13
Solvay Conference 21, 73
special relativity 8, 177
Spinoza 49
statistics 109, 132
Sudarshan, E. G. C. 27
Tantra 75
theodicy 48
thought experiments 73,
  108
time 92, 126, 131, 170,
  183
topos 101
transcategorial unity 93
uncertainty 54, 128
unified field theory 13,
  91
Vaiseshika 78
vibrations 62, 75
Vico, Giambattista 101
Vienna Circle 6
Wenyuan, Yao 158
Wertheimer, M. 29, 169
Westcott, M. R. 30
Williams, W. C. 119, 122,
  137
wonder 85
world government 145
yoga 88
Zedong, Mao 151, 154

About the Editor and Contributors

KEITH R. BURICH is Associate Dean of Arts and Sciences at Canisius College in Buffalo, New York. He has recently published articles on Henry Adams and the Second Law of Thermodynamics, and on science and the crisis of faith among American intellectuals at the end of the nineteenth century. He is currently working on the impact of randomness and indeterminacy on the American mind at the turn of the century.

RICHARD DOBRIN was Professor of Physics at Sarah Lawrence College, Bronxville, New York, at the time of Hofstra's Einstein conference. He has experience in applied physics in the medical electronics field.

CAROL DONLEY is Associate Professor of English at Hiram College, Hiram, Ohio. She has an abiding research interest in the interrelationships of science and literature, especially in the twentieth century.

RICHARD FLEMING is Professor of Philosophy at Bucknell University. He is the author of several articles including, "Einstein's Concept of Rationality in Religion and Science," and "A Literary Understanding of Philosophy: Remarks on the Spirit of Cavell's *The Claim of Reason*." His current research includes work on Interdisciplinary Study and Cognition, and on Wittgenstein and Language Ontology. He is an associate editor of the journal of the College English Association, *The CEA Critic*.

KENNETH FOX is Professor of Physics and Astronomy at the University of Tennessee and Discipline Scientist at the National Aeronautics and Space Administration. He has written extensively on astrophysics in scientific journals and books, and has lectured and taught courses in special and general relativity, as well as in science and public policy. His present responsibilities include management of the Planetary Atmospheres Program at NASA headquarters.

EDWARD FRIEDMAN is Professor of Political Science at the University of Wisconsin at Madison. He is the author of a forthcoming book, *Chinese Village, Socialist State*. He has long been interested in political theory in China and its interrelationships with the philosophy of science.

CHARLES A. FRYE is Professor and Core Coordinator in Philosophy at the Banneker Honors College at Prairie View A&M University. As the former Chair of Black Studies at Hampshire College, he founded *The New England Journal of Black Studies*. His major publications include *Values in Conflict: Blacks and the American Ambivalence Toward Violence*, and *Towards a Philosophy of Black Studies*.

ASHOK K. GANGADEAN is Professor of Philosophy at Haverford College. With a research background in logical theory, philosophy of language, ontology, and Indian and comparative philosophy, he has been increasingly concerned with exploring the problems of the unity of rationality and the holistic nature of human understanding and reality. He has published several research papers on diverse topics developing the method of universal hermeneutics and the logic of meditative reason.

WILLIAM GERTZOG is a former student from the Department of Education, Cornell University, Ithaca, N.Y.

IRVING HAUPTMAN was, until his recent retirement, Professor of Physics and Physical Science at the State University of New York at Fashion Institute of Technology.

ROBERT HAUPTMAN is Assistant Professor of library science at St. Cloud State University. He is the author of *The Pathological Vision: Jean Genet, Louis-Ferdinand Celine, & Tennessee Williams*, and is currently working on a monograph entitled, *Ethical Problems in Librarianship*.

PETER HEWSON was a visiting professor at the Department of Education, Cornell University, Ithaca, N.Y., at the time of Hofstra's Einstein Centennial Symposium. He is now Associate Professor of curriculum and instruction at the University of Wisconsin at Madison.

PATRICK J. HURLEY is Professor of Philosophy at the University of San Diego. He is the author of *A Concise Introduction to Logic*, now in its second edition, and of *Time in the Earlier and Later Whitehead*. He also contributed a paper to *Physics and the Ultimate Significance of Time*, ed. David R. Griffin. His Ph.D. is from St. Louis University.

STANLEY L. JAKI is Professor of Physics at Seton Hall University and a historian of science. His contribution to this volume represents the opening address to Hofstra University's Einstein Centennial Symposium.

WILLIAM F. LAWHEAD is Assistant Professor of Philosophy and Religions at the University of Mississippi at Oxford and previously taught at Trinity College in Illinois. His research has been concerned with the philosophy of language and religious epistemology.

DOROTHY MICHELSON LIVINGSTON has no academic degrees nor honorary diplomas. She was educated mostly by her parents. Her father taught her to paint, play the violin, sail a boat, and play tennis. Her mother gave her an interest in botany, biology, and languages. She speaks French and German fluently, Spanish and Italian, enough to communicate. She is the author of a biography of Michelson entitled, *The Master of Light*. She saw Albert Einstein frequently when he visited her father in Pasadena in 1931.

STEPHEN R. MANDELL is Associate Professor of English at Drexel University. Although his principal research interest is the teaching of writing, he has a strong secondary interest in the interaction of science and modern American poetry. He is co-author of the widely known texts, *Patterns for College Writing* and *Writing: A College Rhetoric*. He is also co-author of the recently published book, *The Holt Handbook*.

WILLIAM MELCZER is Professor of Comparative Literature at Syracuse University. His main interest lies in the history of ideas of the Renaissance and in the Christian iconography of the Byzantine and Romanesque periods. He has published some fifty articles, as well as the volume *La porta di bronzo di Barisano da Trani a Ravello*.

ROY D. MORRISON II is Chairperson of the Master of Theological Studies degree program and Professor of Philosophical Theology, Religion, and Scientific Method, and Philosophy of Black Culture and Religion at Wesley Theological Seminary in Washington, D.C. He previously held a dual appointment as an Assistant Professor on the faculties of the Divinity School and of the New Collegiate Division at the University of Chicago. He has published a number of articles in his fields of specialization.

GEORGE POSNER is Associate Professor of Education at Cornell University, Ithaca, N.Y. His interests are in curriculum design and instruction. He is the author of *Course Design*.

THOMAS RENNA is Professor of History at Saginaw Valley State College in Michigan. His research deals largely with medieval intellectual history, particularly monastic. He is the author of *Church and State in Medieval Europe* and numerous studies of political theory, war and peace, Cistercian thought, and literary views of the city.

ROBERT I. REYNOLDS is Assistant Professor of Psychology at Fordham University at Lincoln Center and formerly held professorships at the Universities of Pennsylvania, Louisville, and other American and foreign institutions. His primary area of research is that of human cognition with a special emphasis upon those factors which underly the creative process.

DENNIS A. ROHATYN is Professor of Philosophy at the University of San Diego. He is the author of *Naturalism and Deontology*, *Two Dogmas of Philosophy*, and a forthcoming book, *The Reluctant Naturalist*, and co-editor of *In Our Own Image*. He also edited *Insight, Prophecy and Moral Vision*, a special double issue of *Cogito* devoted to George Orwell. His Ph.D. is from Fordham University.

JOSEPH R. ROYCE is Research Professor at the Center for Advanced Study in Theoretical Psychology, University of Alberta. He founded the Center in 1967 and was its Director until 1983. He has published over one hundred articles and ten books, primarily in the areas of theoretical psychology, experimental psychology, and behavioral genetics. His books include *Theory of Personality and Individual Differences* (with A. D. Powell), *Theoretical Advances in Behavior Genetics* (with L. P. Mos), and *Multivariate Analysis and Psychological Theory*.

DENNIS P. RYAN is Assistant Professor of Chemistry at Hofstra University. His research interests center on biophysical chemistry, particularly the study of the interaction of small molecules with DNA.

KENNETH STRIKE is Professor of Education at Cornell University, Ithaca, N.Y. His interests are in political philosophy and the philosophy of science. He is the author of a forthcoming book, *The Ethics of Public School Administration*.

BURTON H. VOORHEES is Professor of Mathematics at Athabasca University. He has published a number of papers on relativity theory, genetics, ecology, and psychology. He is currently conducting research in neural modeling and personality.

Hofstra University's
Cultural and Intercultural Studies
Coordinating Editor, Alexej Ugrinsky

George Sand Papers: Conference Proceedings, 1976
(Editorial Board: Natalie Datlof, Edwin L. Dunbaugh, Frank S. Lambasa,
Gabrielle Savet, William S. Shiver, Alex Szogyi)

George Sand Papers: Conference Proceedings, 1978
(Editorial Board: Natalie Datlof, Edwin L. Dunbaugh, Frank S. Lambasa,
Gabrielle Savet, William S. Shiver, Alex Szogyi)

Heinrich von Kleist Studies
(Editorial Board: Alexej Ugrinsky, Frederick J. Churchill, Frank S. Lambasa,
Robert F. von Berg)

William Cullen Bryant Studies
(Editors: Stanley Brodwin, Michael D'Innocenzo)

*Walt Whitman: Here and Now
(Editor: Joann P. Krieg)

*Harry S. Truman: The Man from Independence
(Editor: William F. Levantrosser)

*Nineteenth-Century Women Writers of the English-Speaking World
(Editor: Rhoda B. Nathan)

*Lessing and the Enlightenment
(Editor: Alexej Ugrinsky)

*Dostoevski and the Human Condition After a Century
(Editors: Alexej Ugrinsky, Frank S. Lambasa, and Valija K. Ozolins)

*The Old and New World Romanticism of Washington Irving
(Editor: Stanley Brodwin)

*Woman as Mediatrix
(Editor: Avriel Goldberger)

---

*Available from Greenwood Press